Excel

Get the Results You Want!



# Year 7 NAPLAN*-style Numeracy Tests

Lyn Baker

PASCAL PRESS

* This is not an officially endorsed publication of the NAPLAN program and is produced by Pascal Press independently of Australian governments.

Reprinted 2011
New NAPLAN Test question formats added 2012
Reprinted 2014, 2016 (twice), 2017, 2018

**Revised in 2020 for the NAPLAN Online tests**

Reprinted 2020, 2021, 2022, 2023, 2024, 2025

ISBN 978 1 74125 361 0

Pascal Press Pty Ltd
PO Box 250
Glebe NSW 2037
www.pascalpress.com.au

Publisher: Vivienne Joannou
Project editor: Mark Dixon
Edited by Rosemary Peers
Answers checked by Peter Little
Cover, page design and typesetting by DiZign Pty Ltd and lj Design
Printed by Vivar Printing/Green Giant Press

# Contents

# NAPLAN AND NAPLAN ONLINE

## WHAT IS NAPLAN?

- NAPLAN stands for National Assessment Program—Literacy and Numeracy.
- It is conducted every year in March and the tests are taken by students in Years 3, 5, 7 and 9.
- The tests cover Literacy—Reading, Writing, Conventions of Language (spelling, grammar and punctuation)—and Numeracy.

## WHAT IS NAPLAN ONLINE?

### Introduction

- In the past all NAPLAN tests were paper tests.
- From 2022 all students have taken the NAPLAN tests online.
- This means students complete the NAPLAN tests on a computer or tablet.

### Tailored test design

- With NAPLAN paper tests, all students in each year level took exactly the same tests.
- In the NAPLAN Online tests this isn't the case; instead, every student takes a tailor-made test based on their ability.
- Please visit the official ACARA site for a detailed explanation of the tailored test process used in NAPLAN Online and also for general information about the tests: https://nap.edu.au/online-assessment.
- These tailor-made tests mean broadly, therefore, that a student who is at a standard level of achievement takes a test mostly comprised of questions of a standard level; a student who is at an intermediate level of achievement takes a test mostly comprised of questions of an intermediate level; and a student who is at an advanced level of achievement takes a test mostly comprised of questions of an advanced level.

### Different question types

- Because of the digital format, NAPLAN Online contains more question types than in the paper tests. In the paper tests there are only multiple-choice and short-answer question types. In NAPLAN Online, however, there are also other question types. For example, students might be asked to drag text across a screen or measure a figure with an online ruler.
- Please refer to the next page to see some examples of these additional question types that are found in NAPLAN Online and how they compare to questions in this book. As you will see, the content tested is exactly the same but the questions are presented differently.

# NAPLAN ONLINE QUESTION TYPES

| Additional NAPLAN Online question types | Equivalent questions in this book |
|---|---|

### Additional NAPLAN Online question types

**Ordering**

Order these fractions from lowest to highest.

| $\frac{1}{3}$ | $\frac{3}{8}$ | $\frac{5}{12}$ | $\frac{7}{24}$ |
|---|---|---|---|
| | | | |

lowest highest

### Equivalent questions in this book

Place these fractions $\frac{1}{3}, \frac{3}{8}, \frac{5}{12}, \frac{7}{24}$ in order from lowest to highest.

| | | | |
|---|---|---|---|
| | | | |

lowest highest

### Additional NAPLAN Online question types

**Matching**

$X$ is the point (2, 2). Four other points are plotted on the Cartesian plane.

Drag the correct coordinates to each point.

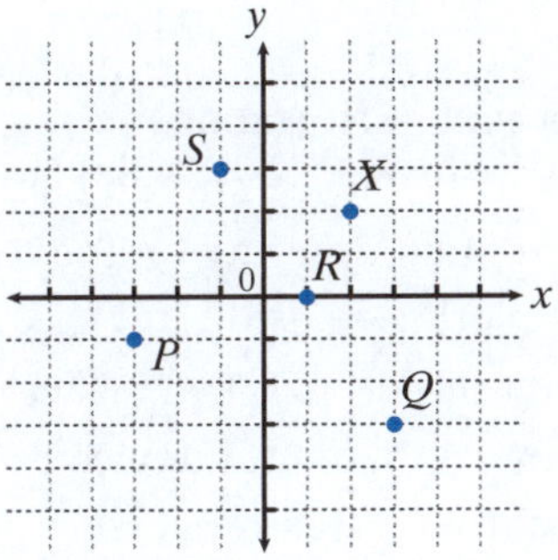


| | |
|---|---|
| (0, 1) | (1, 0) |
| (−1, 3) | (−3, 1) |
| (−1, −3) | (−3, −1) |
| (3, −3) | (−3, 3) |

### Equivalent questions in this book

$X$ is the point (2, 2). Write the coordinates for the points $P$, $Q$, $R$ and $S$.

Choose from the following:

(0, 1), (1, 0), (−1, 3), (−3, 1), (−1, −3), (−3, −1), (3, −3) and (−3, 3).

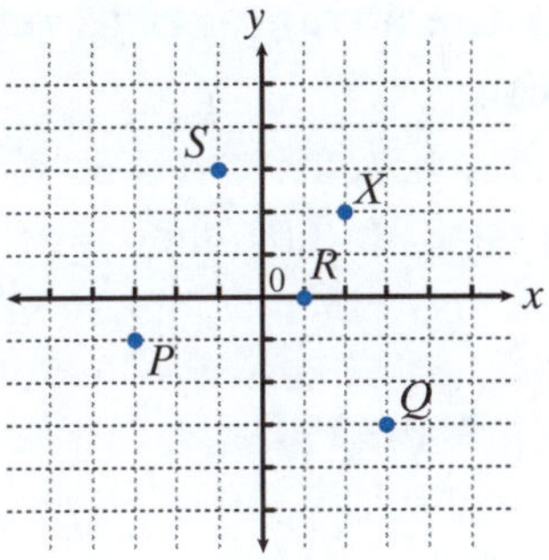


$P$ ______ $Q$ ______

$R$ ______ $S$ ______

### Additional NAPLAN Online question types

**Online ruler**

How long is the nail in the picture below?

Use the online ruler to measure the length of the nail.

______ mm

### Equivalent questions in this book

What is the length of the nail in millimetres?

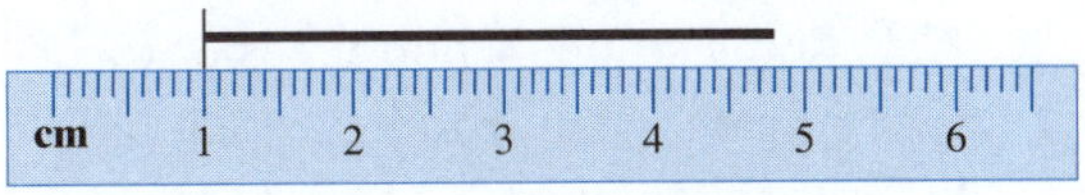


______ mm

# MAXIMISE YOUR RESULTS IN NAPLAN ONLINE

## STEP 1: USE THIS BOOK

### How *Excel* can help you prepare for NAPLAN Online

**Tailored test design**

- We can't replicate the digital experience in book form and offer you tailored tests, but with this series we do provide Standard, Intermediate and Advanced NAPLAN Online–style Numeracy tests—both mini tests and the longer sample tests.
- This means that a student using these tests will be able to prepare with confidence for tests at different ability levels.
- This makes it excellent preparation for the tailored NAPLAN Online Numeracy tests.

### Remember the advantages of revising in book form

There are many benefits to a child using books to prepare for the online test:

- One of the most important benefits is that writing on paper will help your child retain information. It can be a very effective way to memorise. High-quality educational research shows that using a keyboard is not as good as note-taking for learning.
- Students will be able to prepare thoroughly for topic revision using books and then practise computer skills easily. They will only succeed with sound knowledge of topics; this requires study and focus. Students will not succeed in tests simply because they know how to answer questions digitally.
- Also, some students find it easier to concentrate when reading a page in a book than when reading on a screen.
- Furthermore it can be more convenient to use a book, especially when a child doesn't have ready access to a digital device.
- You can be confident that ***Excel*** books will help students acquire the topic knowledge they need, as we have over 30 years experience in helping students prepare for tests. All our writers are experienced educators.

## STEP 2: PRACTISE ON *Excel Test Zone*

### How *Excel Test Zone* can help you practise online

We recommend you go to www.exceltestzone.com.au and register for practice in NAPLAN Online–style tests once you have completed this book. The reasons include:

- for optimal performance in the NAPLAN Online tests we recommend students gain practice at completing online tests as well as completing revision in book form
- students should practise answering questions on a digital device to become confident with this process
- students will be able to practise tailored tests like those in NAPLAN Online, as well as other types of tests
- students will also be able to gain valuable practice in onscreen skills such as dragging and dropping answers, using an online ruler to measure figures and using an online protractor to measure angles.

Remember that ***Excel Test Zone*** has been helping students prepare for NAPLAN since 2009; in fact we had NAPLAN online questions even before NAPLAN tests went online!

We also have updated our website along with our book range to ensure your preparation for NAPLAN Online is 100% up to date.

# ABOUT THE NUMERACY TEST AND THIS BOOK

## THE YEAR 7 NAPLAN ONLINE NUMERACY TEST

### About the test

- This is a Mathematics test in which there are 48 questions.
- Students have 65 minutes to complete the test.

### About the report

- When your child completes the NAPLAN tests, you and your child's school will receive a report indicating their standard of proficiency in literacy and numeracy. There are four levels of achievement:
  - **Exceeding** (advanced proficiency)
  - **Strong** (average to high-average proficiency)
  - **Developing** (not yet proficient)
  - **Needs additional support** (help is needed).

The report will also show the national average.

## ABOUT THIS BOOK

### The Mini Numeracy Tests

- In the first part of the book you will find 32 tests.

  They are divided into three levels of difficulty:
  - Standard level
  - Intermediate level
  - Advanced level.
- The key topics are covered within each level, each with a full page of exercises for your child to practise.
- You will be able to see what level your child is at by finding the point where they start having consistent difficulty with questions. For example, if your child answers most questions correctly up to the intermediate level and then gets most questions wrong from then onwards, it is likely your child's ability is at an intermediate level.
- You will be able to see your child's strengths and weaknesses in different topics by completing the Strengths and weaknesses chart (see page viii).
- You will also be able to give your child intensive practice in short tests which have time limits based on the actual NAPLAN Online Numeracy test times.
- There are quick answers for every question so you can easily mark your child's work. Worked solutions are also included in case you need to show your child how to do a question.

### The Sample Online-style Numeracy Tests

- In the second part of the book we provide you with four sample tests.
- Your child will be able to practise doing these longer sample tests which are modelled on actual NAPLAN Online Numeracy tests.
- The sample tests are at different levels and consist of:
  - one standard-level test
  - two intermediate-level tests
  - one advanced-level test.

# STRENGTHS AND WEAKNESSES CHART

- As the student completes each test, mark it using the answer section at the back and then fill in this chart to record their progress.
- You will be able to see at a glance the student's strengths and weaknesses in different topics of Numeracy.
- If the student needs more practice on specific topics, they can use the checklist of ***Excel*** books on the back cover to find the book to help them.

| Type of test | Level | Mini test | Topic | Mark |
|---|---|---|---|---|
| Non-calculator | Standard | 1 | Numbers | /24 |
| Non-calculator | Standard | 2 | Patterns | /15 |
| Calculator Allowed | Standard | 3 | Geometry | /16 |
| Calculator Allowed | Standard | 4 | Numbers and Patterns | /23 |
| Calculator Allowed | Standard | 5 | Measurement | /17 |
| Calculator Allowed | Standard | 6 | Tables and Graphs | /16 |
| Non-calculator | Standard | 7 | Mixed Questions | /16 |
| Calculator Allowed | Standard | 8 | Mixed Questions | /16 |
| Non-calculator | Intermediate | 9 | Fractions, Decimals and Percentages | /24 |
| Calculator Allowed | Intermediate | 10 | Angles and Measurements | /17 |
| Calculator Allowed | Intermediate | 11 | Chance and Data | /11 |
| Calculator Allowed | Intermediate | 12 | Number | /24 |
| Calculator Allowed | Intermediate | 13 | Problems | /16 |
| Calculator Allowed | Intermediate | 14 | Patterns | /15 |
| Calculator Allowed | Intermediate | 15 | Time | /17 |
| Calculator Allowed | Intermediate | 16 | Measurement | /16 |
| Calculator Allowed | Intermediate | 17 | Perimeter, Area and Volume | /12 |
| Calculator Allowed | Intermediate | 18 | Ratio and Rates | /20 |
| Non-calculator | Intermediate | 19 | Mixed Questions | /16 |
| Calculator Allowed | Intermediate | 20 | Mixed Questions | /16 |
| Non-calculator | Advanced | 21 | Fractions and Percentages | /16 |
| Non-calculator | Advanced | 22 | Decimals, Negatives and Ratio | /20 |
| Non-calculator | Advanced | 23 | Patterns and Algebra | /14 |
| Calculator Allowed | Advanced | 24 | Space | /12 |
| Calculator Allowed | Advanced | 25 | Number | /16 |
| Calculator Allowed | Advanced | 26 | Patterns and Algebra | /12 |
| Calculator Allowed | Advanced | 27 | Perimeter, Area and Volume | /16 |
| Calculator Allowed | Advanced | 28 | Angles | /16 |
| Calculator Allowed | Advanced | 29 | Rates and Measurements | /14 |
| Calculator Allowed | Advanced | 30 | Chance and Data | /11 |
| Non-calculator | Advanced | 31 | Mixed Questions | /17 |
| Calculator Allowed | Advanced | 32 | Mixed Questions | /14 |

NON-CALCULATOR Standard level questions

20 MIN

# Mini Test 1: Numbers

1 $(6 \times 1000) + (4 \times 10) + 7 = ?$

**A** 647 **B** 6407 **C** 6047 **D** 6470

2 $3 \times \square = 6 \times 5$

What is the missing number? ☐

3 What number is **five thousand and twenty-eight**?

**A** 5028 **B** 5208 **C** 5082 **D** 500 028

4 Certain packets of nails each hold 50 nails. William has 20 packets.
How many nails does he have? ☐

5 Which is **not** equal to 12?

**A** $4 \times 3$ **B** $5 + 7$ **C** $20 - 8$ **D** $18 \div 3$

6 $\square = 3$ and $\otimes = 5$

$\square + \otimes = \triangle + \triangle$

$\triangle = ?$

**A** 3 **B** 4 **C** 5 **D** 6

7 The arrow points to a position on the number line.

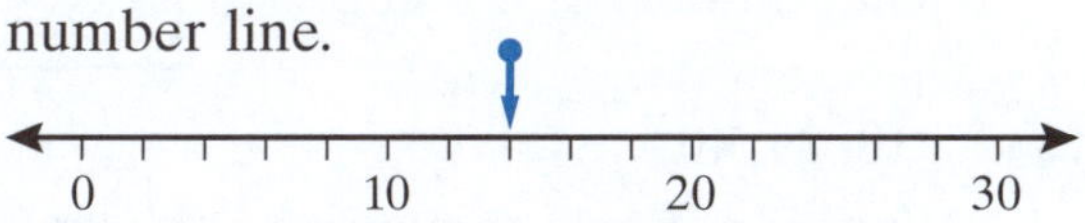


What is the number at this position?

**A** 12 **B** 13 **C** 14 **D** 15

8 Kayleen bought a packet of 24 stickers. She has used 15. How many stickers are left? ☐

9 Which is **not** equal to 16?

**A** $20 - 3 + 1$ **B** $3 \times 5 + 1$
**C** $5 + 9 + 2$ **D** $6 \times 3 - 2$

10 Kiah has $7.80. She needs 45 cents more to have enough to buy a particular book.
How much is the book? $ ☐

11 Which is another way of writing $4^3$?

**A** $3 + 3 + 3 + 3$ **B** $4 + 4 + 4$
**C** $3 \times 3 \times 3 \times 3$ **D** $4 \times 4 \times 4$

12 $0.2 + 0.02 =$ ☐

13 Leo divided 186 by a number and got 1.86 as the answer.
What number did Leo divide by? ☐

14 Look at these cards.

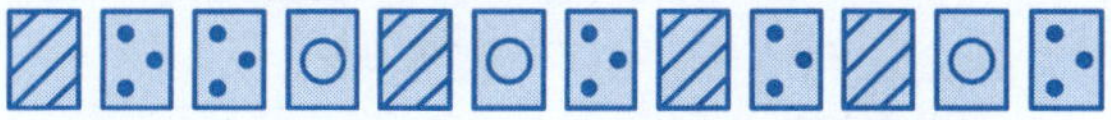

What fraction of the cards has stripes?

**A** $\frac{1}{2}$ **B** $\frac{1}{3}$ **C** $\frac{1}{4}$ **D** $\frac{1}{6}$

15 Which has the same value as $8 \times 4$?
Select **all** correct answers.

**A** $10 \times 2 + 3$ **B** $10 \times 3 + 2$
**C** $10 \times 4 - 8$ **D** $10 \times 2 \times 3$

16 $4228 \div 7 = ?$

**A** 64 **B** 604 **C** 640 **D** 6004

17 George paid for an ice-cream with a $5 note. He received $2.75 in change.
How much was the ice-cream? $ ☐

18 In July 2010 the population of Pinkton was 215 684. What is this population rounded to the nearest thousand?

**A** 200 000 **B** 220 000
**C** 216 000 **D** 215 000

19 Which diagram does **not** have $\frac{1}{4}$ shaded?

20 $5 - 0.4 =$ ☐

21 Which is equal to $17 \times 35 + 83 \times 35$?

**A** $100 \times 35$ **B** $17 \times 48$
**C** $17 \times 118$ **D** $83 \times 52$

22 In a box of 40 pens, 20% are red. How many pens are red?

**A** 2 **B** 4 **C** 6 **D** 8

23 Walter bought 4 kg of watermelon. The watermelon was selling for 79 cents per kilogram. Which is closest to the amount Walter paid?

**A** $1 **B** $2 **C** $3 **D** $4

24 The population of Yakata is 3.5 million. The population of Zenin is 8.2 million.
What is the difference between the populations? ☐ million



Answers and explanations on pages 86–87

NON-CALCULATOR Standard level questions

## Mini Test 2: Patterns

**1** What number comes next in this pattern?

25 50 75 100 125 ☐

**2** What is the rule to continue this number pattern?

127, 119, 111, 103, 95, …

**A** increase by 7 **B** decrease by 7
**C** increase by 8 **D** decrease by 8

**3** Candy used the rule 'multiply the previous number by 2 and then add 3' to get the next number in her pattern. If the third number was 11, what was the fourth number? ☐

**4** One of the second numbers is missing in this table.

| **First number** | 1 | 2 | 3 | 4 | 8 |
|---|---|---|---|---|---|
| **Second number** | 2 | 4 | 6 | 8 | ? |

What number is missing? ☐

**5** Isaac is making a pattern of figures with matches.

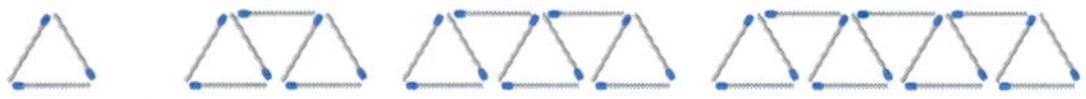

Figure 1 Figure 2 Figure 3 Figure 4

How many matches will Isaac need to make Figure 5? ☐

**6** What is the next number in this pattern?

993, 995, 997, 999, ☐

**7** Rose is shading boxes in a counting pattern beginning with 3.

| | | | | | | | | | |
|---|---|---|---|---|---|---|---|---|---|
| 1 | 2 | **3** | 4 | 5 | 6 | 7 | 8 | 9 | **10** |
| 11 | 12 | 13 | 14 | 15 | 16 | **17** | 18 | 19 | 20 |
| 21 | 22 | 23 | **24** | 25 | 26 | 27 | 28 | 29 | 30 |
| 31 | 32 | 33 | 34 | 35 | 36 | 37 | 38 | 39 | 40 |

The last number that Rose shaded was 24. What is the next number that Rose should shade?

**A** 31 **B** 33 **C** 35 **D** 37

**8** What is the next number in this pattern?

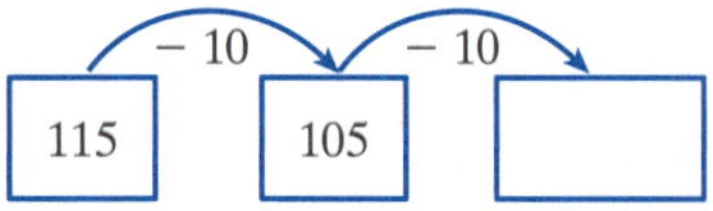


**9** Ming has written down the first five numbers in a pattern: 1, 2, 4, 8, 16

What is the seventh number in the pattern? ☐

**10** Sallyanne is making a pattern with pins.

Shape 1 Shape 2 Shape 3

How many pins will she need for the fifth shape? ☐

**11** Raad wrote down the first six numbers in a pattern: 1, 2, 4, 7, 11, 16

What is the next number in Raad's pattern? ☐

**12** A number is missing in this table.

| **First number** | 1 | 2 | 3 | 4 | 5 |
|---|---|---|---|---|---|
| **Second number** | 7 | 11 | 15 | 19 | ? |

What number is missing? ☐

**13** 7, 18, 29, 40, …

What is the rule to continue this number pattern?

**A** increase by 9 **B** increase by 11
**C** increase by 19 **D** increase by 21

**14** Precious used the rule 'add 6 to the previous number' to get the next number in her pattern. The first number in Precious's pattern was 5. What was the 10th number?

**A** 51 **B** 56 **C** 59 **D** 65

**15** Tea has been spilled on this table and only one of the second numbers can be read.

| **First number** | 1 | 2 | 3 | 4 | 5 |
|---|---|---|---|---|---|
| **Second number** | | | | 7 | |

Which could be the rule to find the second number? Select **all** possible answers.

**A** Add 3 to the first number.
**B** Subtract the first number from 10.
**C** Double the first number and subtract 1.
**D** Multiply the first number by 3 and add 4.


Answers and explanations on pages 87–89

CALCULATOR ALLOWED | **Standard level questions**

# Mini Test 3: Geometry

15 MIN

**1** This 3D object is a

**A** cone. **B** cube.
**C** cylinder. **D** sphere.

**2** This 2D shape is a

**A** pentagon. **B** parallelogram.
**C** rhombus. **D** trapezium.

**3** Chloe folded this net to make a cube.

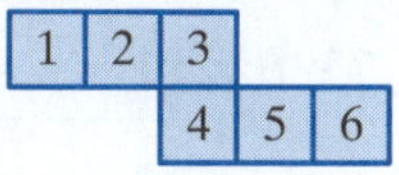


What number is on the face opposite the face numbered 3?

**4** Which 3D object can be made from this net?

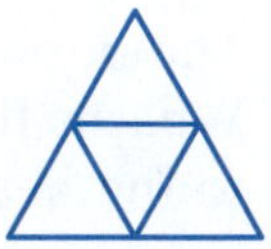

**A** triangular prism
**B** triangular pyramid
**C** hexahedron
**D** octahedron

**5** Which shape does not have any parallel sides?

One of the corners was cut from a rectangular prism as shown in the diagram.

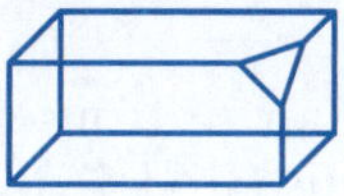

Use this information to answer questions 6 to 8.

**6** How many faces does the shape now have?

**7** How many edges does the shape now have?

**8** How many vertices does the shape now have?

**9** Which diagram shows an angle of about 45°?

**10** Which shape has exactly two lines of symmetry?

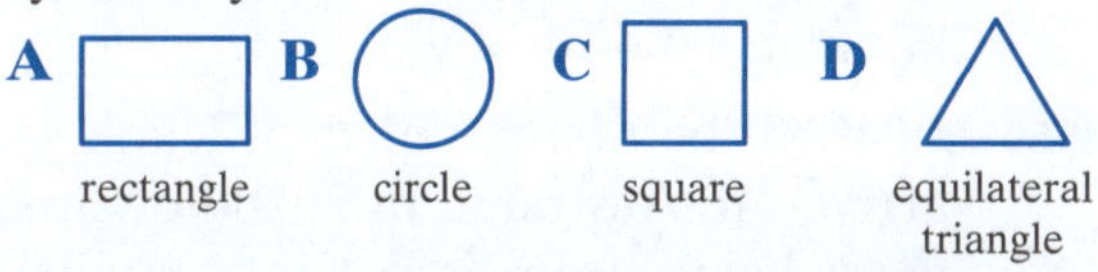


This object is made from ten cubes. Use this information to answer questions 11 to 13.

**11** What is the view from the top?

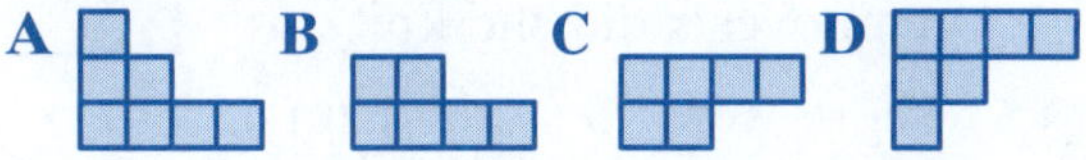


**12** What is the view from the right side?

**13** What is the view from the back?

Use this diagram to answer questions 14 to 16.

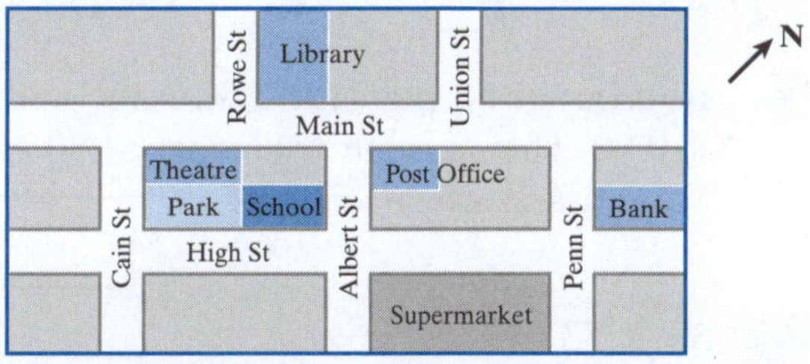


**14** Which best describes the direction of the school from the post office?

**A** South-West **B** South
**C** South-East **D** East

**15** Gavin leaves the bank, travelling along High Street and turns into Albert Street. He turns left into Main Street and then takes the next street on the right. What street is Gavin now in?

**A** Cain **B** Rowe **C** Union **D** Penn

**16** Serafina is standing in the park facing High Street. She turns through a quarter turn in a clockwise direction. What is she now facing?

**A** High Street **B** school
**C** theatre **D** Cain Street


Answers and explanations on pages 89–90

CALCULATOR ALLOWED **Standard level questions**

20 MIN

# Mini Test 4: Numbers and Patterns

**1** Ibuki paid $30 for potatoes at $2.50 per kilogram. How many kilograms did he buy? ☐ kg

**2** In 1885 the first member of Kim's family arrived in Australia. In 2020 how many years has it been since Kim's family arrived in Australia?

**A** 165 **B** 155 **C** 145 **D** 135

**3** Sue is paid $24 for every vest she knits. If she is paid $768, how many vests did she knit? ☐

**4** There were 25 students in a class. The average mark in an exam was 76. What was the total of all the marks? ☐

**5** 5% of all houses in a town have a swimming pool. If there are 2600 houses in the town, how many pools are there? ☐

**6** What fraction of 24 is 18?

**A** $\frac{2}{3}$ **B** $\frac{3}{4}$ **C** $\frac{3}{2}$ **D** $\frac{4}{3}$

**7** Ryo bought 3 pies and 2 sausage rolls for $10.80. The sausage rolls cost $1.80 each. How much was each pie? $ ☐

**8** 38 + 53 = 81 + ☐

What number will make this number sentence true? ☐

**9** Carmel divided 6534 by a number and the answer was 198. What number did she divide by? ☐

**10** The calculation $\frac{786 - 528}{32 - 26}$ gives the cost per hour, (in dollars), of hiring a contractor. What is the cost per hour? $ ☐

**11** Today a tanker collected 8250 litres of milk from a dairy farm. This was an increase of 370 litres on the previous collection two days ago. How many litres did the tanker collect two days ago? ☐ litres

**12** Ellie bought sheets priced at $49.95 and a pillow. She received $12.75 change from $100. How much was the pillow? $ ☐

**13** Which fraction has the same value as $3\frac{2}{5}$?

**A** $\frac{11}{5}$ **B** $\frac{17}{5}$ **C** $\frac{19}{5}$ **D** $\frac{21}{5}$

**14** What is the next number in this pattern?

88, 75, 62, 49, ☐

**15** △ = 2 and □ = 5

△ + ○ = □ − △

○ =

**A** 1 **B** 2 **C** 3 **D** 4

**16** The population of a city is 7 512 862. What is this number rounded to the nearest thousand? ☐

**17** Which of these has the same value as 9 × 7? Select **all** correct answers.

**A** 12 + 9 × 3 **B** 11 × 6 − 3

**C** 8 × 8 − 3 **D** 12 × 5 + 3

**18** Which of these is between $4^2$ and $6^2$?

**A** 5 **B** 14 **C** 31 **D** 45

**19** Each person paid $8.50 to watch a movie. If the total takings were $544, how many people watched the movie? ☐

**20** To make a pair of trousers, Harry bought 2.5 m of material at $6.00 per metre. How much did he pay? $ ☐

**21** What is the next number in this pattern?

3, 3.6, 4.2, 4.8, ☐

**22** One-third of all animals at a shelter are cats. There are 57 cats at the shelter. How many animals are at the shelter? ☐

**23** Two-thirds of students taking a course were male. If there were 816 students taking the course, how many were female? ☐



Answers and explanations on pages 90–91

CALCULATOR ALLOWED **Standard level questions**

# **Mini Test 5:** Measurement

**1** Rhianna travelled 300 km in 5 hours. What was her average speed in kilometres per hour?

**A** 15 **B** 60 **C** 75 **D** 150

**2** A scale painted on the wall of a pool measures the depth of water in the pool.

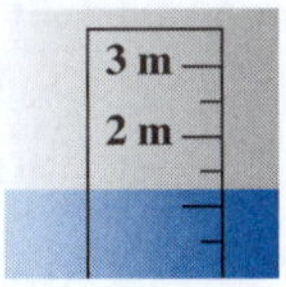


About how deep is the water?

**A** 2.25 m **B** 2.75 m **C** 1.25 m **D** 1.75 m

**3** The shaded area on the grid, in square units, is closest to

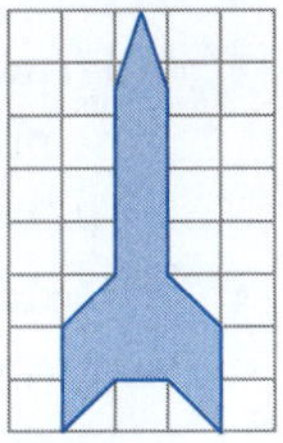

**A** 7 **B** 10
**C** 12 **D** 13

**4** What is the length of the nail in millimetres?

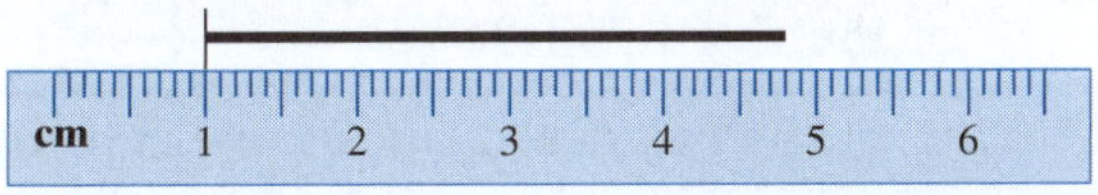


☐ mm

**5** Which calculation will give the area of the shaded triangle?

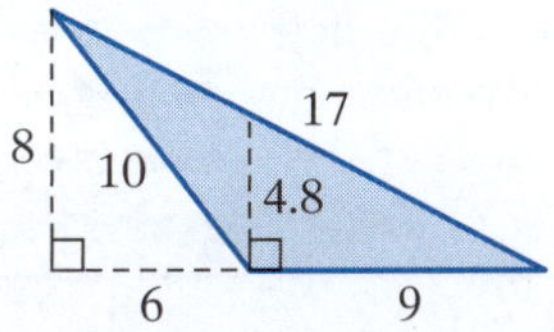


**A** $\frac{1}{2} \times 9 \times 8$ **B** $\frac{1}{2} \times 9 \times 10$

**C** $\frac{1}{2} \times 9 \times 4.8$ **D** $\frac{1}{2} \times 9 \times 17$

**6** What is the length of each side of a square that has perimeter 36 cm? ☐ cm

**7** Samuel arrives at the station at a quarter past two. His train leaves at 2:51. How many minutes must Samuel wait for the train? ☐ min

**8** 7.25 kg is equal to

**A** 72 500 g **B** 725 000 g
**C** 725 g **D** 7250 g

**9** Which jug has the most juice?

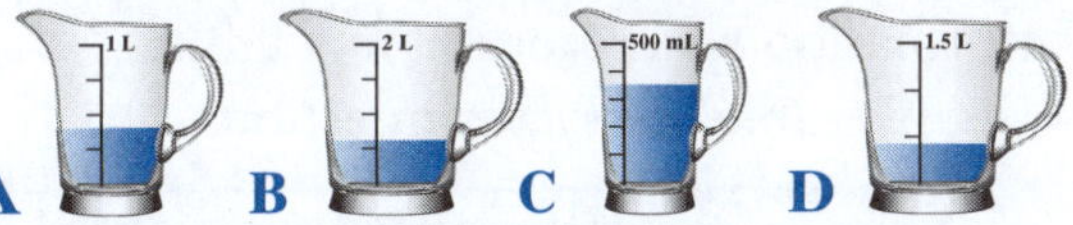


**A** **B** **C** **D**

**10** A rectangular paddock is 100 m long and 50 m wide. What is the area?

**A** 1500 $m^2$ **B** 300 $m^2$
**C** 500 $m^2$ **D** 5000 $m^2$

**11** When Jim went to bed his bedside clock showed 18:15.

What time did Jim go to bed?

**A** 6:15 pm **B** 8:15 pm
**C** 9:15 pm **D** 10:15 pm

**12** Daniel rides 8 km in half an hour. What is his average speed in kilometres per hour?

**A** 4 **B** 12 **C** 16 **D** 24

**13** What is the perimeter of this rectangle?

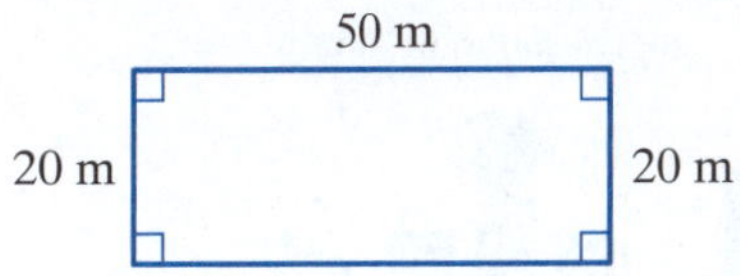


☐ m

**14** A car uses petrol at the rate of 9 L/100 km. How many litres will it use to travel 500 km?

**A** 45 **B** 56 **C** 450 **D** 560

**15** How many centimetres are in 50 m? ☐ cm

**16** Hannah catches a train at 8:40 am and arrives at her destination at 9:15 am. How many minutes long is the journey? ☐ min

**17** One-third of the length of a fence is 60 m. How long is the fence? ☐ m


Answers and explanations on pages 91–92

CALCULATOR ALLOWED | Standard level questions

# Mini Test 6: Tables and Graphs

Some boys and girls were asked whether or not they had ever travelled overseas and the information was recorded in a table.

Use this table to answer questions 1 to 4.

| | Had | Had not | Total |
|---|---|---|---|
| **Boys** | 23 | 61 | |
| **Girls** | 19 | | 56 |
| **Total** | | | |

**1** How many girls had travelled overseas?

**2** How many girls had not travelled overseas?

**3** How many fewer boys had travelled overseas than had not travelled overseas?

**4** In total, how many boys and girls were surveyed?

The graph shows the number of mistakes made by students in a quiz. Use the graph to answer questions 5 to 8.

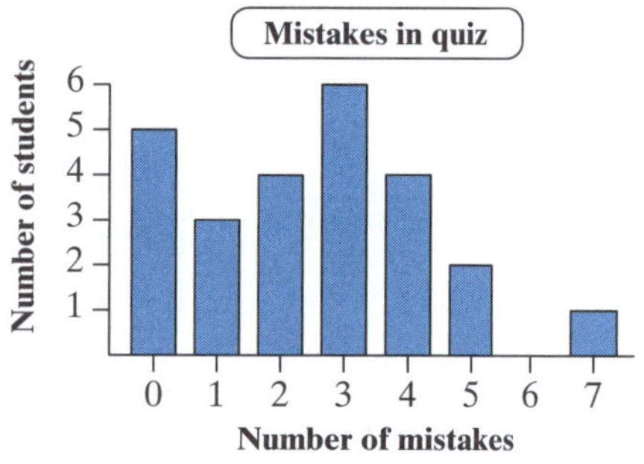


**5** How many students made no mistakes?

**6** What is the most common number of mistakes?

**A** 3 **B** 4 **C** 5 **D** 6 **E** 7

**7** How many more students made 4 mistakes than 2 mistakes?

**A** 0 **B** 1 **C** 2 **D** 3

**8** Which **two** statements are correct?

**A** The greatest number of mistakes was 7.
**B** No student made exactly 6 mistakes.
**C** 12 students made fewer than 2 mistakes.
**D** 5 people made exactly 2 mistakes.

Use this table to answer questions 9 to 12.

| Town | Population |
|---|---|
| Amity | 8700 |
| Bunyip | 6550 |
| Carrington | 4350 |
| Douglass | 5125 |
| Earlwood | 4800 |

**9** How much greater is the population of Bunyip than that of Earlwood?

**10** How much less is the population of Douglass than that of Amity?

**11** How much more is the combined population of Carrington and Douglass than the population of Bunyip?

**12** What is the total population of the five towns?

Use this graph to answer questions 13 to 16.

**13** How many cars were serviced on Wednesday?

**14** How many more cars were serviced on Friday than on Monday?

**15** How many cars were serviced last week?

**16** Which statement is **not** correct?

**A** 21 cars were serviced on Thursday.
**B** The same number of cars were serviced on Monday and Thursday as the number on Tuesday and Friday.
**C** One more car was serviced on Wednesday than on Friday.
**D** A total of 32 cars were serviced on Thursday and Friday.




Answers and explanations on pages 93–94

# Mini Test 7: Mixed Questions

1. Write the number five hundred and six thousand and twenty.

2. In which diagram is both the height and width of this shape doubled?

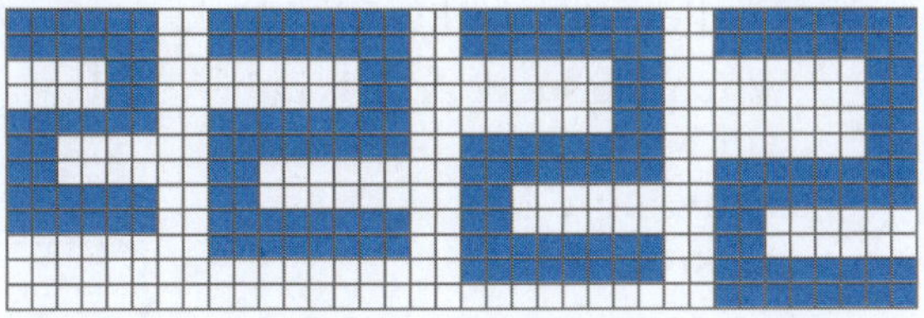

A B C D

3. Which diagram shows this shape after it has made a quarter turn clockwise?

A B C D

4. Guy drove 200 km in 4 hours. What was his average speed (in kilometres per hour)?
**A** 50 **B** 60 **C** 70 **D** 80

5. The table shows the number of shirts sold in three different sizes and three different sleeve styles.

| Sleeve type | Small | Medium | Large |
|---|---|---|---|
| Sleeveless | 23 | 15 | 8 |
| Short | 34 | 29 | 28 |
| Long | 21 | 24 | 14 |

How many more small sleeveless shirts were sold than long-sleeved large shirts?

6. $(5 \times 1000) + (2 \times 100) + 6 = ?$
**A** 526 **B** 5260 **C** 5206 **D** 5026

7. Place these decimals in order from the lowest to the highest: 0.4, 0.07, 0.019, 0.25

8. Which of these does **not** have the same number of faces as vertices?

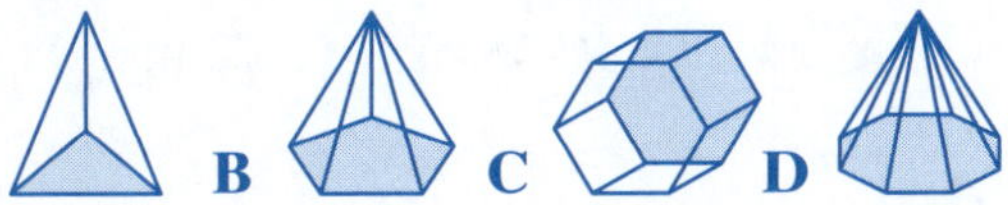

A B C D

9. Which dotted line is a line of symmetry?

A B C D

10. What is the perimeter of this shape?

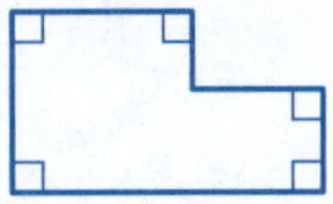


m

11. What is the next number in this pattern?
15, 23, 31, 39,

12. $32 \div 8 = 8 \div \otimes$
What is $\otimes$?

13. A group of women were asked how many children they had had. The results are shown in the graph.

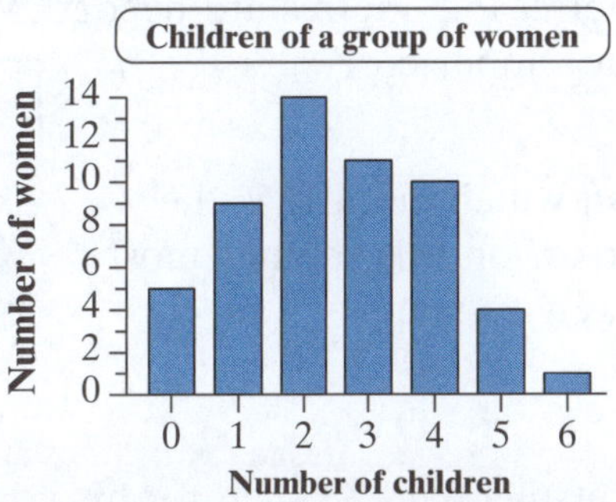


How many more women had had three children than five children?

14. Of which solid could this be the net?

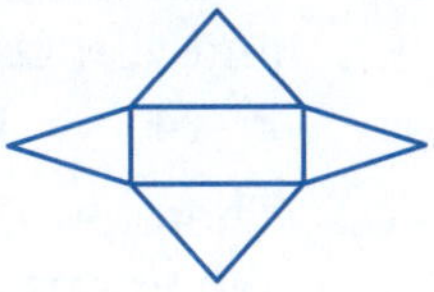

**A** triangular prism
**B** triangular pyramid
**C** rectangular prism
**D** rectangular pyramid

15. What is the average (mean) of the numbers 3, 5 and 10?

16. Which is another way of writing $3^2$?
**A** $3 \times 2$ **B** $2 + 2 + 2$
**C** $3 \times 3$ **D** $2 \times 2 \times 2$


Answers and explanations on pages 94–96

CALCULATOR ALLOWED

Standard level questions

# Mini Test 8: Mixed Questions

**1** Which arrow points to the position of 0.85 on this number line?

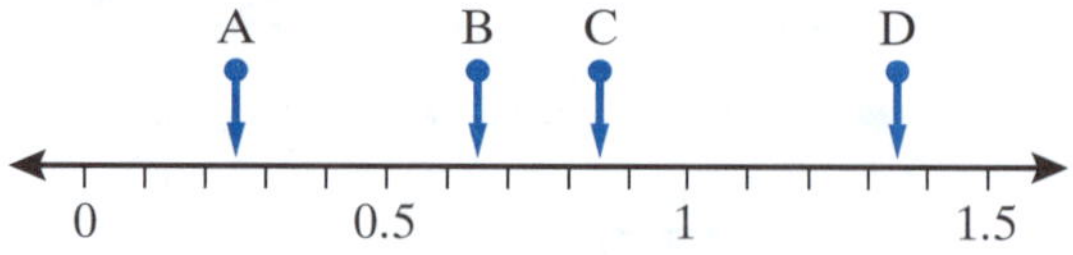


**A** A **B** B **C** C **D** D

**2** Which of these shapes is a pentagon?

**3** This graph shows the results when 800 people voted for 5 different candidates in an election. About how many voted for Joe?

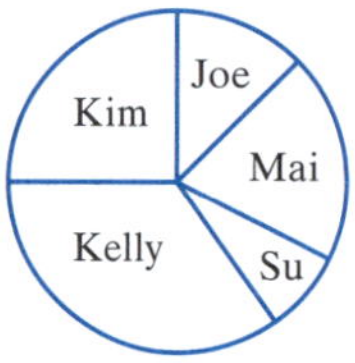


**A** 200 **B** 160 **C** 100 **D** 50

**4** The timber for a railing costs $36.40 per metre. 15.5 metres are needed. What will the timber cost?

$ ☐

**5** On which number is the arrow on this spinner most likely to stop?

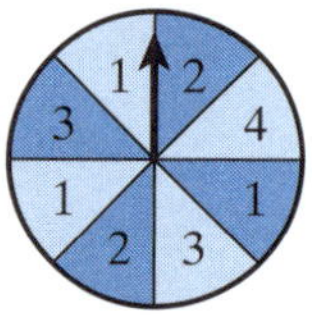


**A** 1 **B** 2
**C** 3 **D** 4

**6** Katsuya paid $3 each for his goats to be tested for a disease. Altogether Katsuya paid $582. How many goats were tested?

☐

**7** What is the total number of faces, edges and vertices of a cube?

**A** 24 **B** 26 **C** 28 **D** 30

**8** There are 24 students in Ms Locke's class. When raising money for a charity, the average amount raised was $7 per student. How much did the class raise for the charity?

$ ☐

**9** Moses bought 32 cans of drink for $27.20. What was the cost per can?

$ ☐

**10** Abigail started to mow the lawn at 4:38 pm and she finished at 5:17 pm. How long did Abigail take to mow the lawn?

**A** 39 min **B** 55 min
**C** 1 h 21 min **D** 1 h 45 min

**11** Ajak is building a cube from smaller identical cubic blocks.

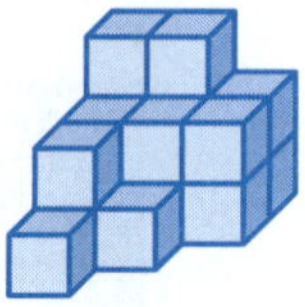

When finished, Ajak's cube will be 4 blocks high. How many more blocks does Ajak need?

☐

**12** The perimeter of a square is 144 m. What is the length of a side of the square?

**A** 36 m **B** 24 m **C** 12 m **D** 6 m

**13** What is the value of $x$ in the diagram?

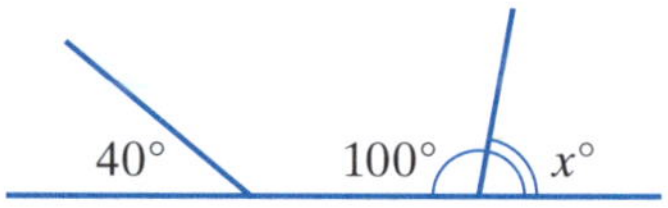


**A** 40 **B** 50 **C** 60 **D** 80

**14** Anna wrote out the first five numbers in a pattern: 700, 622, 544, 466, 388

What is the next number in this pattern?

☐

**15** A survey was taken of 180 spectators as they entered the ground to watch a football match. Each of the spectators was asked which team they expected to win the match. Some of the results are shown in the table.

| | Birds | Serpents |
|---|---|---|
| **Males** | 38 | 69 |
| **Females** | 27 | |

How many females thought the Serpents would win?

☐

**16** What is the area of this triangle?

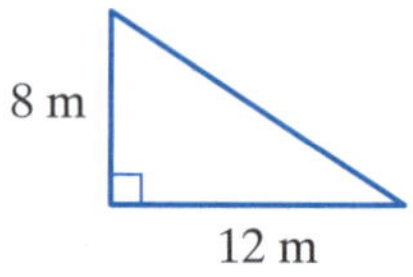


**A** 24 $m^2$ **B** 48 $m^2$ **C** 72 $m^2$ **D** 96 $m^2$



Answers and explanations on pages 96–97

NON-CALCULATOR **Intermediate level questions**

## Mini Test 9: Fractions, Decimals and Percentages

**1** What is 15% of 240?

**2** Jenny gave $12 000 to a charity's winter collection. This was 30% of the total collection. How much was the total collection?

$

**3** What number is halfway between 0.35 and 0.53?

**4** What number comes next in this pattern?

0.7, 0.64, 0.58, 0.52,

**5** What is $\frac{2}{3} + \frac{3}{5}$?

**A** $\frac{5}{8}$ **B** $\frac{2}{5}$ **C** $1\frac{4}{15}$ **D** $1\frac{1}{6}$

**6** Garth's petrol bill last month was $240. This month it was $300. What percentage increase is this?

**A** 20% **B** 25% **C** $33\frac{1}{3}$% **D** 40%

**7** $\frac{2}{3}$ of a number is 12. What is the number?

**A** 8 **B** 9 **C** 16 **D** 18

**8** $11 + 1.1 + 0.11 = ?$

**A** 12.21 **B** 12.12 **C** 11.22 **D** 11.3

**9** What is $8 \div \frac{1}{2}$?

**A** 4 **B** 6 **C** 12 **D** 16

**10** Place the numbers 0.63, 0.6003, 0.6033 and 0.603 in order from the lowest to the highest.

**11** What is the average of 0.2, 0.7 and 0.9?

**12** Which are the best approximations for 0.576 and 4.6183?

**A** 0.5 and 4.6 **B** 0.5 and 4.7
**C** 0.6 and 4.6 **D** 0.6 and 4.7

**13** $1\frac{1}{2} \times 1\frac{1}{3} = ?$

**A** $1\frac{1}{6}$ **B** $1\frac{5}{6}$ **C** 2 **D** $2\frac{1}{6}$

**14** Which of $\frac{1}{2}$, 0.5 or 50% is the largest?

**A** $\frac{1}{2}$ **B** 0.5
**C** 50% **D** They are all equal.

**15** Jesse had 2.5 kg of flour in a canister. He took 0.25 kg from the canister and then added another 0.5 kg to the canister. How much flour was then in the canister?

kg

**16** $3 - 1\frac{3}{4} = ?$

**A** $\frac{1}{4}$ **B** $1\frac{1}{4}$ **C** $1\frac{3}{4}$ **D** $2\frac{1}{4}$

**17** As a member of staff, Precious was able to pay 20% less than the original price when she bought a television, originally priced at $900. How much did Precious pay?

$

**18** Which is equivalent to $\frac{2}{5}$?

**A** $\frac{2+4}{5+4}$ **B** $\frac{2 \times 4}{5 \times 4}$ **C** $\frac{2^2}{5^2}$ **D** $\frac{10 \div 2}{10 \div 5}$

**19** $6 \div 0.3 = ?$

**A** 0.02 **B** 0.2 **C** 2 **D** 20

**20** $\frac{3}{10} + \frac{17}{100} = ?$

**A** 0.47 **B** 0.317 **C** 0.173 **D** 0.2

**21** Which is the largest?

**A** $\frac{2}{3}$ **B** $\frac{7}{12}$ **C** $\frac{5}{6}$ **D** $\frac{3}{4}$

**22** What is the value of $\frac{12 + 6}{2 \times 3}$?

**A** 3 **B** 12 **C** 27 **D** 45

**23** 70% of workers at a factory are male. If there are 80 workers altogether, how many are female?

**24** $187 \times 34 = 6358$

What is $18.7 \times 3.4$?



Answers and explanations on pages 97–99

CALCULATOR ALLOWED

Intermediate level questions

# Mini Test 10: Angles and Measurements

**1** What is the size of this angle?

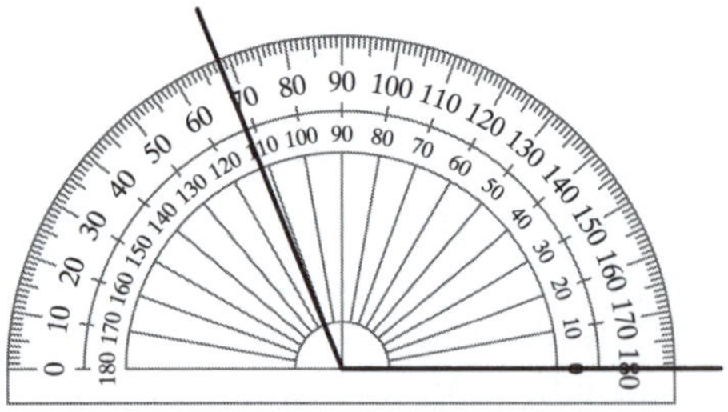

°

**2** Which of these angles is reflex?

**3** How many 250 mL glasses can be filled from a 3 L bottle of juice?

**4** An angle of 110° would be

**A** acute. **B** reflex.
**C** obtuse. **D** complementary.

**5**

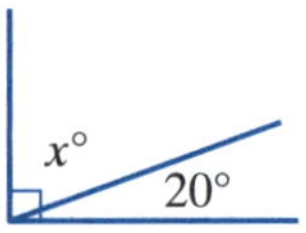


$x =$

**6** How many minutes are in $10\frac{3}{4}$ hours?

**7** Which is the same as 1 tonne and 200 kilograms? Select **all** correct answers.

**A** 1.2 t **B** 1.02 t
**C** 1.002 t **D** 1 000 200 kg
**E** 10 200 kg **F** 1200 kg

**8** Which is the best estimate for the size of this angle?

**A** 30° **B** 70° **C** 120° **D** 170°

**9** How many square centimetres are in 1 square metre?

**A** 100 **B** 1000 **C** 10 000 **D** 100 000

**10** *ABCD* is a straight line.

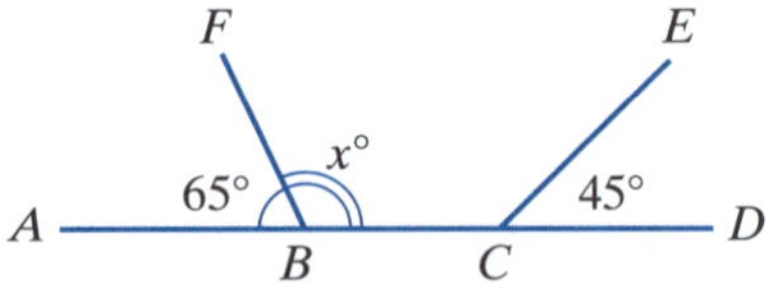


$x =$

**11** Which angle would measure about 50°?

**12** *EC* and *DB* are straight lines.
What is the size of the shaded angle?

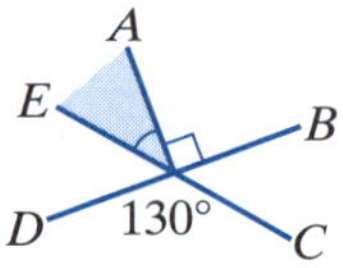


°

**13**

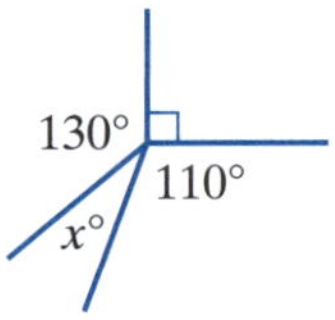


$x =$

**14** This diagram is made up of a square and an equilateral triangle.

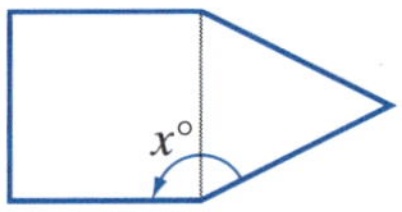


$x =$

**15** An isosceles triangle is cut along its axis of symmetry. The two resulting triangles **must** be

**A** right-angled. **B** isosceles.
**C** equilateral. **D** scalene.

**16** A megalitre is one million litres. How many kilolitres are in 5 megalitres?

**17** Place these lengths in order from the shortest to the longest:

0.003 km, 7 m, 4000 cm, 20 000 mm



Answers and explanations on pages 99–100

CALCULATOR ALLOWED **Intermediate level questions** 15 MIN

# Mini Test 11: Chance and Data

**1** There are 30 pegs in a bag. 12 are red and the others are green. Clint takes a peg from the bag without looking. What is the chance of his picking a red peg?

**A** $\frac{1}{3}$ **B** $\frac{2}{3}$ **C** $\frac{2}{5}$ **D** $\frac{3}{5}$

**2** Winter found the average (mean) of ten scores: 7, 7, 7, 7, 8, 8, 8, 9, 9, 10. When a single score is left out, the average of the remaining nine scores is 8. Which score is left out?

**A** 7 **B** 8 **C** 9 **D** 10

**3** 150 boys and girls were asked if they had seen a particular movie. Some of the results are shown in the table.

| | Seen | Not seen | Total |
|---|---|---|---|
| **Boys** | 31 | — | — |
| **Girls** | 17 | 54 | 71 |
| **Total** | 48 | — | 150 |

How many boys hadn't seen the movie? ☐

**4** A bag holds 11 blue, 17 green, 9 yellow and 8 red marbles. Without looking, Mark takes a marble from the bag. What is the chance that the marble is yellow?

**A** $\frac{1}{6}$ **B** $\frac{1}{5}$ **C** $\frac{1}{4}$ **D** $\frac{1}{3}$

**5** This graph can be used to change distances in miles to distances in kilometres.

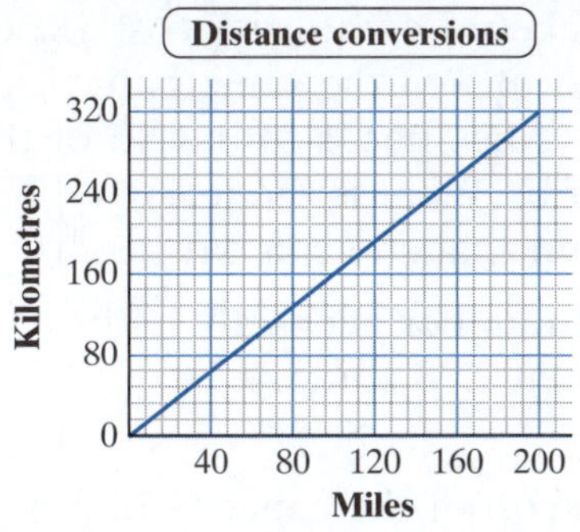


It is 180 miles from Ayr to Brady. About what distance is this in kilometres?

**A** 120 km **B** 180 km
**C** 220 km **D** 280 km

**6** Henning's average mark after his first five tests was 81. His average mark after his first six tests was 82. What did Henning score in his sixth test? ☐

**7** Dimple threw two dice together and got a total of 12.

She then threw the dice again. In how many ways can she get a total of 5?

**A** 2 **B** 3 **C** 4 **D** 5

**8** The table shows the amount of time Katherine spent travelling to and from work this week.

| Day | Time |
|---|---|
| Monday | 45 min |
| Tuesday | $1\frac{1}{4}$ h |
| Wednesday | 40 min |
| Thursday | $1\frac{1}{2}$ h |
| Friday | 1 h 10 min |

Complete this sentence:

The average time Katherine spent travelling per day is 1 hour and ☐ minutes.

**9** This spinner has sections in 4 different colours; red, blue, green and yellow. It is to be spun 24 times.

Which colour would you expect to get 8 times?

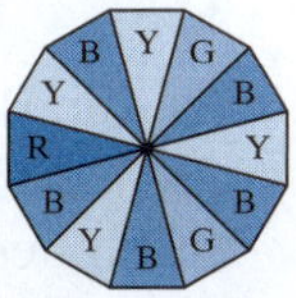


**A** red **B** blue
**C** green **D** yellow

**10** In which bucket is the chance of choosing blue one in six?

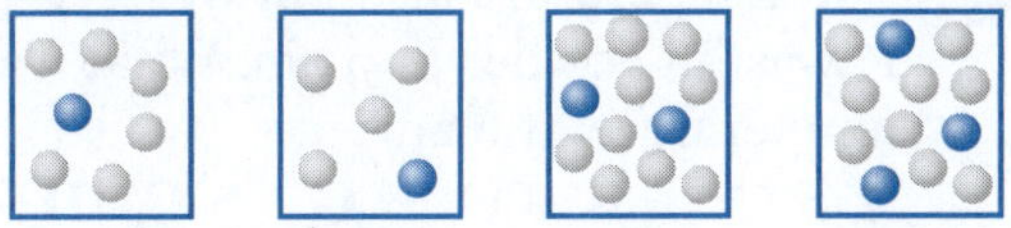

**A** **B** **C** **D**

**11** The table shows the number of students who scored different marks in a quiz.

| **Mark** | 5 | 6 | 7 | 8 | 9 | 10 |
|---|---|---|---|---|---|---|
| **Number** | 8 | 12 | 17 | 15 | 11 | 9 |

How many more students scored more than 7 than scored less than 7 marks? ☐


Answers and explanations on pages 100–102

CALCULATOR ALLOWED **Intermediate level questions** 25 MIN

# Mini Test 12: Number

**1** Which is equal to $10^2 - 6^2$?
**A** $4^2$ **B** $5^2$ **C** $7^2$ **D** $8^2$

**2** The cost of 8 cans of dog food is \$21.60. What is the cost of 3 cans of dog food? \$ ☐

**3** Which expression is equal to $39 \times 27 - 19 \times 39$?
**A** $39 \times 8$ **B** $27 \times 19$
**C** $20 \times 12$ **D** $58 \times 39$

**4** Which is the best approximation to the value of $\frac{3}{7}$ as a decimal?
**A** 0.37 **B** 0.38 **C** 0.42 **D** 0.43

**5** Bob paid \$6.30 for grapes. The grapes were selling for \$8.40 a kilogram. How many kilograms did Bob buy? ☐ kg

**6** What number is the arrow pointing to on this number line?

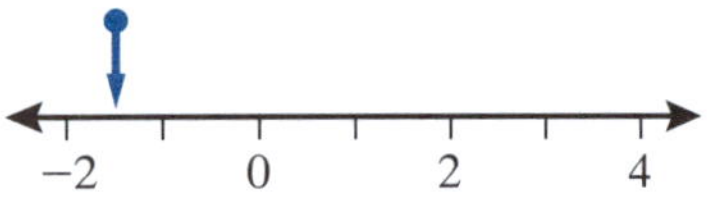


☐

**7** Which numbers are prime? Select **all** correct answers.
**A** 51 **B** 53 **C** 55 **D** 57 **E** 59

**8** To find the answer to a question, Shelly needed to divide by 7 but instead she multiplied by 7. Shelly's answer was 490. What should it have been?
**A** 10 **B** 70 **C** 3430 **D** 24 010

**9** The temperature at 6 pm was −2°C. It was 7°C colder at 6 am. What was the temperature at 6 am?
**A** 5°C **B** 9°C **C** −5°C **D** −9°C

**10** $3.26 + 4.83 = 5.17 + \square$
What number will make this number sentence true? ☐

**11** What is the value of $\frac{147.85 - 37.29}{0.08}$? ☐

**12** A positive number is multiplied by itself and then 17 is subtracted. The answer is 8. What was the number? ☐

**13** What two numbers multiply to 24 and add to 11? ☐ and ☐

**14** A box holds 24 chocolates. 16 of the chocolates have cream centres. What fraction of the chocolates has cream centres?
**A** $\frac{1}{3}$ **B** $\frac{2}{3}$ **C** $\frac{1}{2}$ **D** $\frac{2}{5}$

**15** Tanvi bought a scooter for \$77, a doll for \$28 and some clothes for \$132. What is the best way to estimate the total cost?
**A** \$80 + \$30 + \$140 **B** \$80 + \$30 + \$130
**C** \$70 + \$30 + \$130 **D** \$70 + \$20 + \$140

**16** Which is equivalent to $2^3 \times 4^2$?
**A** $2 \times 3 \times 4 \times 2$ **B** $2 \times 2 \times 4 \times 4$
**C** $2 \times 4 \times 4 \times 4$ **D** $2 \times 2 \times 2 \times 4 \times 2$

**17** Which number has both 6 and 8 as factors?
**A** 18 **B** 24 **C** 36 **D** 64

**18** $-2 + 5 \times 2 \square 6 - 3 \times 2$
What symbol can be placed in the box to make the statement true?
**A** $<$ **B** $=$ **C** $>$ **D** $||$

**19** $\frac{2}{3}$ of a number is 24. The number is
**A** 16 **B** 26 **C** 36 **D** 46

**20** $\sqrt{25 + 144} =$
**A** 13 **B** 14 **C** 16 **D** 17

**21** A packet of 4 batteries cost \$8.30 and a packet of 3 of the same batteries cost \$6.40. Alice needs to buy 15 of the batteries. What is the least amount that she can pay? \$ ☐

**22** What number is exactly halfway between 2.78 and 5.46? ☐

**23** One-sixth of the capacity of a bottle is 240 mL. What is the capacity of the bottle? ☐ mL

**24** Last year 200 people had a particular disease. This year 120 people had the disease. What percentage decrease is this?
**A** 33% **B** 40% **C** 60% **D** 66%


Answers and explanations on pages 102–103

CALCULATOR ALLOWED **Intermediate level questions** 20 MIN

# Mini Test 13: Problems

**1** 9 cans of drink cost $8.55.
What would 5 of the same cans cost? $ ☐

**2** Nikita bought a packet of 45 balloons. She shared them with her friends. Nikita and her friends had 6 balloons each and there were 3 left over. With how many friends did Nikita share the balloons?
**A** 5 **B** 6 **C** 7 **D** 8

**3** When 6 is added to a certain number and the result multiplied by 7 the answer is 133. What is that number? ☐

**4** Erik put 5 marbles in each of 8 bags and had 2 marbles left over. If he were to put 6 marbles in each of the bags, how many more marbles would he need? ☐

**5** Corinne spends $\frac{1}{8}$ of her monthly income on rent. If Corinne's rent is $1320 per month, how much is her **yearly** income? $ ☐

**6** Johanna scored these marks in her exams.

| Subject | Mark |
|---|---|
| English | 81 out of 100 |
| Maths | 39 out of 50 |
| Science | 32 out of 40 |
| Art | 45 out of 60 |

Arrange the subjects in order from the highest percentage mark to the lowest percentage mark.
☐

**7** Which onions are cheapest?
**A** 4 kg for $4.20 **B** 5 kg for $4.90
**C** 6 kg for $5.70 **D** 3 kg for $3.15

**8** Together a ruler and a pen cost $2.90. The pen cost $1.30 more than the ruler. How much was the ruler? $ ☐

**9** Sarah spends $\frac{1}{3}$ of her weekly income on food and $\frac{1}{5}$ on rent. If her weekly income is $1605, how much does Sarah have to spend on other things? $ ☐

**10** Lara bought a car. The cash price of the car was $9500 but Lara didn't have that much cash. She paid $2000 to begin with and agreed to pay $700 every month for 12 months. How much extra will Lara pay by paying off the car instead of paying cash? $ ☐

**11** Jerome has a total of 36 cars and trucks. He has three times as many trucks as cars. How many trucks does Jerome have? ☐

**12** Jim is 3 years older than Pete but 2 years younger than Luke. The sum of all their ages is 62. How old is Luke? ☐

**13** Sergio arranges the plants he has for sale in 8 rows with 18 plants in each row. He decides to rearrange the plants into 12 rows. How many plants will there be in each row? ☐

**14** The maximum temperature on Monday was 15 °C. The minimum temperature on Tuesday was 21° colder than that. What was the minimum temperature on Tuesday? ☐ °C

**15** There are 140 adults and children at a fete. The ratio of adults to children is 2 to 5. How many children are at the fete?
**A** 40 **B** 56 **C** 84 **D** 100

**16** Hayley's weekly bus ticket costs $11.20. A daily return ticket costs $2.70. How much does Hayley save by buying a weekly ticket instead of 5 return tickets? $ ☐


Answers and explanations on pages 103–105

CALCULATOR ALLOWED

Intermediate level questions

20 MIN

# Mini Test 14: Patterns

**1** Daniel wrote out the first four numbers in a pattern: 3, 7, 11, 15, …

What is the 10th number in the pattern? ☐

**2** △ and ☐ stand for numbers that are related by a rule.

| △ | 2 | 3 | 4 | 5 | 6 |
|---|---|---|---|---|---|
| ☐ | 4 | 7 | 10 | 13 | 16 |

What is the rule?

**A** ☐ = 2 × △ **B** ☐ = △ + 2

**C** ☐ = 3 × △ − 2 **D** ☐ = 4 × △ − 4

**3** Freya wrote down the first four numbers of a pattern that began with 111 and subtracted 7 each time. Which is the correct pattern?

**A** 111, 103, 96, 89 **B** 111, 104, 96, 89

**C** 111, 103, 97, 90 **D** 111, 104, 97, 90

**4** What is the seventh number in a pattern that begins with 6 and goes up by 9 each time? ☐

**5** Sally wrote down some of the numbers in a pattern but the fifth number was missing.

1, 3, 6, 10, **?**, 21, 28, 36

What is the missing number? ☐

**6** This pattern of pentagons is made with matchsticks.

Pentagons

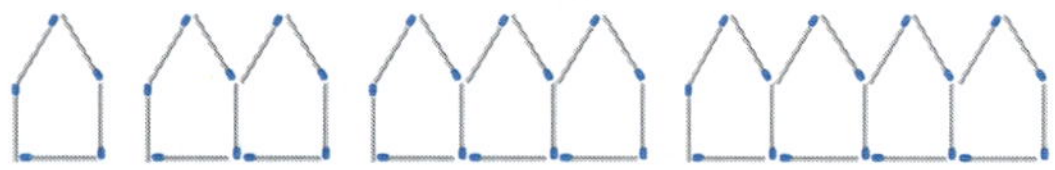

| Number of pentagons | 1 | 2 | 3 | 4 |
|---|---|---|---|---|
| Number of matchsticks | 5 | 9 | 13 | 17 |

How many matchsticks would be needed to make 6 pentagons? ☐

**7** A table has been drawn up to show the relationship between numbers in a pattern.

| First number | 1 | 2 | 3 | 4 | 5 |
|---|---|---|---|---|---|
| Second number | 5 | 11 | ? | 23 | 29 |

What is the missing number?

**A** 15 **B** 17 **C** 19 **D** 21

**8** What is the next number in this pattern?

$$\frac{1}{5}, \frac{1}{4}, \frac{3}{10}, \frac{7}{20}, \ldots$$

**A** $\frac{2}{5}$ **B** $\frac{1}{2}$ **C** $\frac{7}{10}$ **D** $\frac{9}{25}$

**9** Oliver wrote out the first four numbers in a pattern: 3, 6, 12, 24.

What is the next number? ☐

**10** This pattern of hexagons is made with pins.

Hexagons

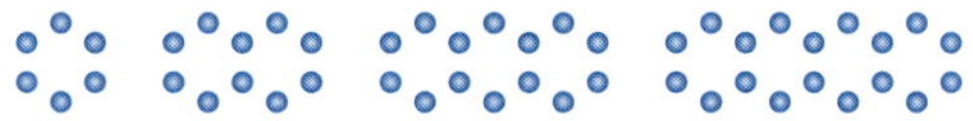

| Number of hexagons | 1 | 2 | 3 | 4 |
|---|---|---|---|---|
| Number of pins | 6 | 10 | 14 | 18 |

How many hexagons are made with 30 pins?

**A** 5 **B** 6 **C** 7 **D** 8

**11** What is the next number in this pattern?

1, 3, 7, 15, 31, ☐

**12** Shinhye wrote out a pattern of letters:

C, B, A, D, C, B, A, D, C, B, A, D, C, B, …

What is the 43rd letter in this pattern?

**A** A **B** B **C** C **D** D

**13** A number pattern begins with 3 and goes up by 2 each time. What is the sum of the first five numbers in this pattern? ☐

**14** Which is correct?

**A** *P* is at (2, −3) and *Q* is at (1, −4).

**B** *P* is at (−3, 2) and *Q* is at (1, −4).

**C** *P* is at (2, −3) and *Q* is at (−4, 1).

**D** *P* is at (−3, 2) and *Q* is at (−4, 1).

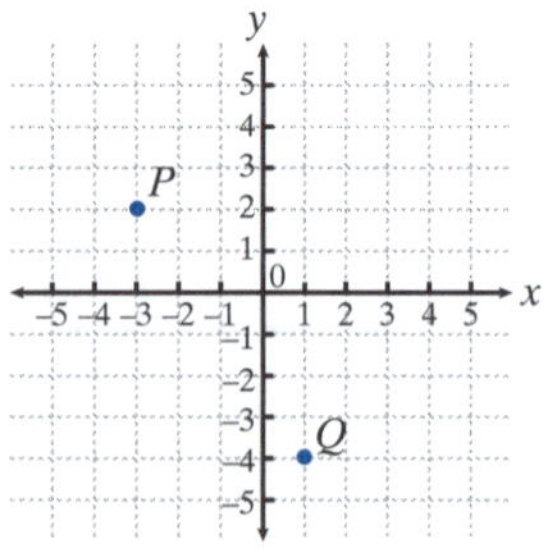


**15** The rule for a pattern is 'double the previous number and add 3'. If the first number in the pattern is 2, what is the third number? ☐



Answers and explanations on pages 105–106

CALCULATOR ALLOWED

Intermediate level questions

# Mini Test 15: Time

**1** How many hours and minutes are between 9:48 am and 9:12 pm on the same day?

**A** 11 h 24 min **B** 11 h 36 min
**C** 12 h 24 min **D** 12 h 36 min

**2** Which time is the same as that shown on this electronic display?

**A** 8:45 pm **B** 9:45 pm
**C** 10:45 pm **D** 11:45 pm

**3** The time in Auckland is 4 hours ahead of the time in Perth. A football match begins in Auckland at 7:30 pm and is shown live on television in Perth. What time does it start in Perth?

**4** When it is 7:30 pm in Hobart it is 9:30 am in Paris on the same day. When it is 6:00 pm on Saturday in Paris, what time is it in Hobart?

**A** 4 am Saturday **B** 8 am Saturday
**C** 4 am Sunday **D** 8 am Sunday

**5** Henry caught a train at 10:46 am. The journey took $1\frac{1}{2}$ hours. What time did Henry reach his destination?

**6** Carrie arrived at the station at quarter to six. The next train was due to leave at 6:25. How many minutes did Carrie have to wait? min

**7** Mei is booked on a cruise that sails from Brisbane on 13 September. She must pay for her ticket at least 21 days in advance. What is the last day that Mei can pay?

**A** 21 August **B** 22 August
**C** 23 August **D** 24 August

**8** If the long minute hand on this clock turns through another 270°, what will be the time?

**A** 7:05 **B** 7:10 **C** 7:15 **D** 7:20

**9** Gisele sees the office clock. Gisele leaves work at 5:15. How many minutes are there before she can leave? min

**10** A plane leaves Launceston at 8:40 am and arrives in Sydney at 10:15 am. How long was the journey?

**A** 1 h 25 min **B** 1 h 35 min
**C** 2 h 25 min **D** 2 h 35 min

**11** If a heart beats 80 times in a minute, about how many times will it beat in a week?

**A** 400 000 **B** 800 000
**C** 2 400 000 **D** 48 000 000

**12** What is the time $4\frac{3}{4}$ hours after 9:40 am?

**A** 1:25 pm **B** 2:15 pm
**C** 2:25 pm **D** 2:55 pm

**13** Nerida arrived at work at 8:40 am and left at 3:15 pm. How long was Nerida at work?

**A** 6 h 25 min **B** 6 h 35 min
**C** 7 h 25 min **D** 7 h 35 min

**14** A train arrived in Sydney at 14:25 after a $2\frac{3}{4}$ hour journey from Newcastle. What time did it leave Newcastle?

**A** 11:10 **B** 11:40 **C** 11:50 **D** 12:10

**15** Which pair of clocks show the same time?

**16** Vera looks at this train timetable.

| | | | | |
|---|---|---|---|---|
| **Lincoln** | 8:14 | 8:56 | 9:27 | 10:06 |
| **Putney** | 8:28 | 9:15 | 9:44 | 10:20 |
| **Zetland** | 8:49 | 9:37 | 10:05 | 10:39 |
| **City** | 9:02 | 9:50 | 10:18 | 10:51 |

What is the latest time that Vera should leave home if it takes her 8 minutes to walk to Putney station and she needs to be in the city before quarter past ten?

**17** 1 hour and 45 minutes is the same as

**A** 175 minutes. **B** 145 minutes.
**C** 1.75 hours. **D** 1.45 hours.




Answers and explanations on pages 106–107

CALCULATOR ALLOWED

Intermediate level questions

# Mini Test 16: Measurement

**1** 1 kg and 60 g is the same as
**A** 1.06 kg. **B** 1.6 kg.
**C** 1.006 kg. **D** 100060 g.

**2** This balance shows that 3 blocks have the same mass as 1 ball.

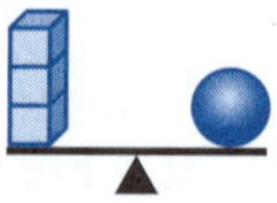

How many blocks would balance 2 balls and a block? [ ]

**3** A piece of farmland is rectangular in shape. It is 800 metres long and 500 metres wide. How many hectares is the farmland?
**A** 4 **B** 40 **C** 400 **D** 4000

**4** 1 cm and 20 mm is the same as
**A** 1.2 cm. **B** 1.02 cm.
**C** 120 mm. **D** 30 mm.

**5** This jug holds juice.
If 500 millilitres of the juice is poured from the jug, how many millilitres remain?

[ ] mL

**6** On a map 1 cm represents 20 km. Two places are 7.8 cm apart on the map. What is the actual distance between the two places? [ ] km

**7** There are several layers in a stack of bricks. There are 12 bricks in each layer. The stack of bricks is 1.8 m high. If each brick is 15 cm high, how many bricks are in the stack?
**A** 144 **B** 180 **C** 216 **D** 225

**8** 1 minute and 15 seconds is the same as
**A** 115 seconds. **B** 125 seconds.
**C** 1.15 minutes. **D** 1.25 minutes.

**9** A block of land is 25 m by 25 m. A scale drawing of the land is shown. What scale is used in the drawing?

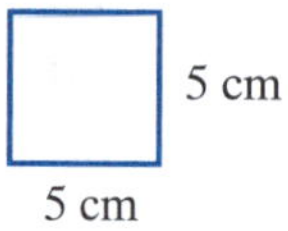


**A** 1 cm represents 5 m.
**B** 1 cm represents 2 m.
**C** 5 cm represents 1 m.
**D** 2 cm represents 1 m.

**10** The diameter of a 20-cent piece is about 28 mm. Kip has $15.00 in 20-cent pieces. If Kip laid out all his 20-cent pieces in a straight line, about how long would the line be?
**A** 1.1 m **B** 2.1 m **C** 3.1 m **D** 4.1 m

**11** 1 metre and 15 centimetres is
**A** 1.015 m. **B** 1015 m.
**C** 1015 cm. **D** 1150 mm.

**12** A photo is 12 cm long and 8 cm wide. It is enlarged so that it is now 54 cm long. How wide is the enlarged photo?
**A** 30 cm **B** 36 cm **C** 45 cm **D** 50 cm

**13** A vat holds 8.75 kilolitres of liquid. How many litres does the vat hold?
**A** 87.5 **B** 875 **C** 8750 **D** 87 500

**14** This balance shows that 2 blocks and a ball have the same mass as 4 balls.

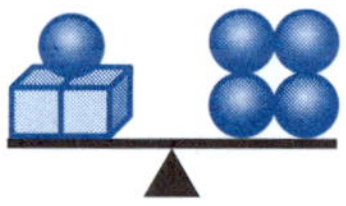

How many blocks would balance 6 balls?
**A** 2 **B** 3 **C** 4 **D** 5

**15** This is a map of Lake Lambert.

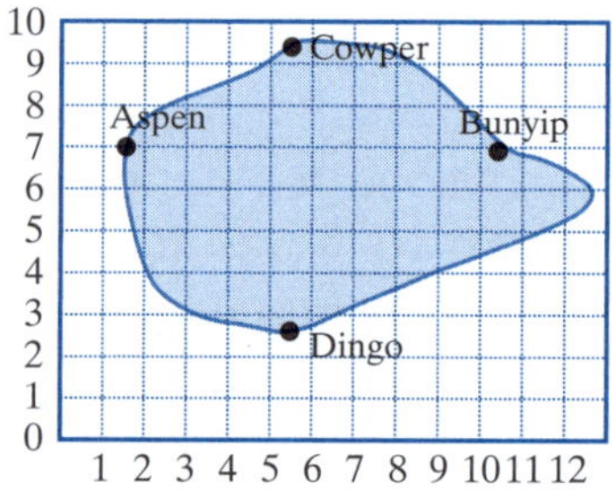


The shortest distance from Aspen to Bunyip is 6.75 km. The shortest distance from Cowper to Dingo is about
**A** 4.75 km. **B** 5.25 km.
**C** 5.75 km. **D** 6.25 km.

**16** Place these in order from the lightest to the heaviest.

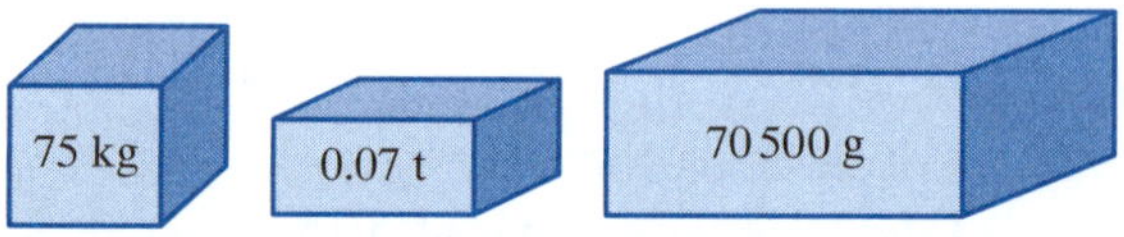


[ ]



Answers and explanations on pages 107–109

CALCULATOR ALLOWED

Intermediate level questions

15 MIN

# Mini Test 17: Perimeter, Area and Volume

**1** All of the faces of this solid cube are to be painted.

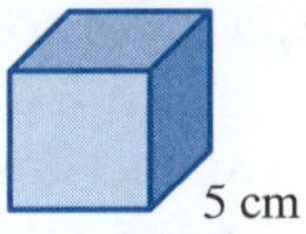


What is the **total** area to be painted? _____ $cm^2$

**2** This shape is made of a rectangle and 7 equilateral triangles. Each side of each triangle is 4 cm long.

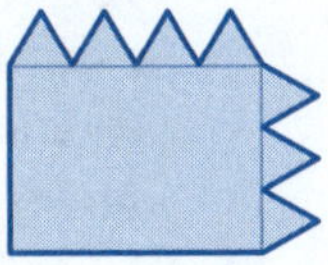

What is the perimeter of the shape? _____ cm

**3** The diagram shows a picture in a rectangular frame. The frame is 12 cm long and 7 cm wide. There is a 1 cm wide border on all sides of the picture.

What is the area of the shaded border?

**A** 18 $cm^2$ **B** 34 $cm^2$
**C** 46 $cm^2$ **D** 66 $cm^2$

**4** What is the perimeter of this shape?

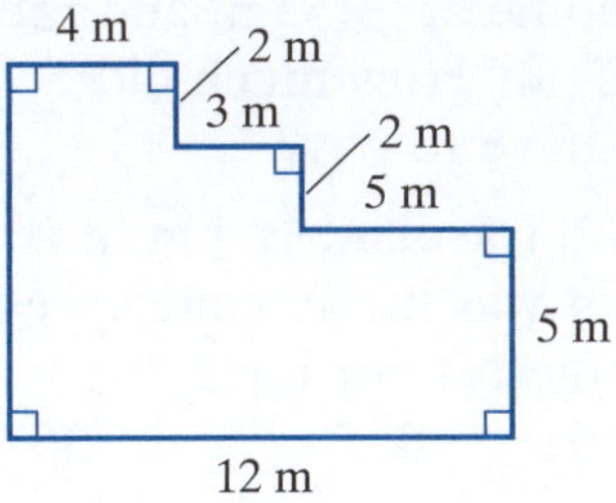


_____ m

**5** The volume of a rectangular prism is 12 cubic metres. Another rectangular prism is twice as long and twice as wide but the same height as the first prism. What is the volume of the second prism (in cubic metres)?

**A** 16 **B** 24 **C** 48 **D** 96

**6** Which paddocks have the same area?

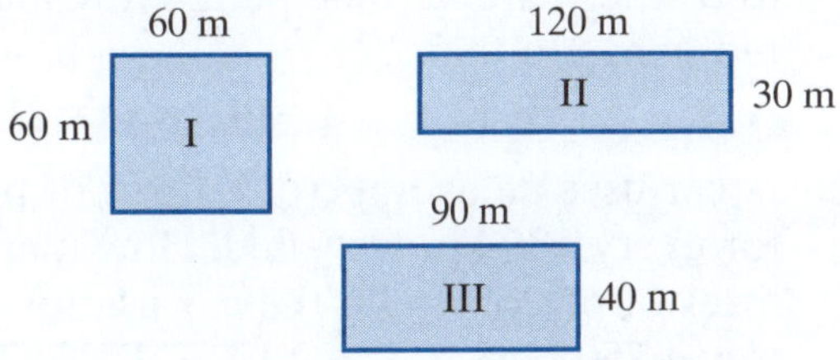


**A** I and II only **B** I and III only
**C** II and III only **D** I, II and III

**7** A rectangle has a perimeter of 32 m. The longer sides are 11 m long. How long are the shorter sides? _____ m

**8** The area of a square is 625 $m^2$. What is the length of each side? _____ m

**9** Which is correct? Select **all** correct answers.

**A** 1 ha = 100 $m^2$ **B** 1 $km^2$ = 100 ha
**C** 1 $m^2$ = 100 $cm^2$ **D** 1 $cm^2$ = 100 $mm^2$

**10** This rectangular yard is to be paved with pavers that are 20 cm long and 10 cm wide.

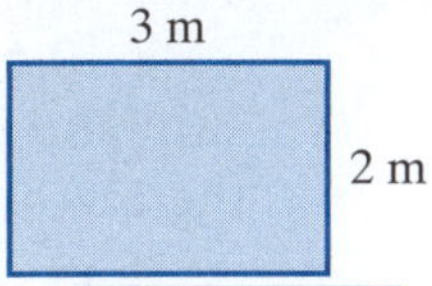


How many pavers will be needed? _____

**11** The perimeter of a square is 36 cm. What is its area? _____ $cm^2$

**12** Place these rectangular prisms in order from the least volume to the greatest volume.

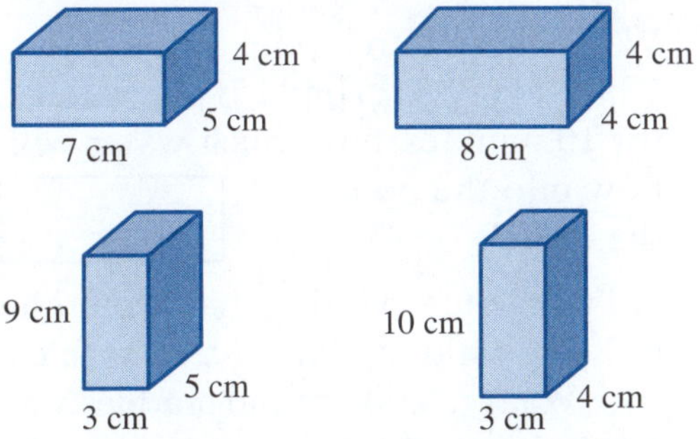


_____


Answers and explanations on pages 109–110

CALCULATOR ALLOWED | Intermediate level questions

25 MIN

# Mini Test 18: Ratio and Rates

**1** The ratio of blue pens to red pens is 7 to 2. There are 42 blue pens. How many red pens are there?

**A** 9 **B** 12 **C** 14 **D** 21

**2** A car uses an average of 8 litres of petrol for every 100 km travelled. How many litres of petrol would the car use to travel 750 km?

☐ litres

**3** Raad ran for two and a half hours. He ran at an average speed of 16 km per hour. How far did he run?

**A** 24 km **B** 36 km **C** 40 km **D** 48 km

**4** There were 180 people at a meeting. The ratio of males to females was 4 to 5. How many females were at the meeting?

**A** 80 **B** 90 **C** 100 **D** 144

**5** A car uses petrol at the rate of 9 litres for every 100 km travelled. How far can the car travel on 54 litres of petrol?

☐ km

**6** Theo knows that he takes 16 paces to walk 10 metres. Theo walks to the bus stop and counts 640 paces. How far is it to the bus stop?

**A** 25 m **B** 40 m **C** 250 m **D** 400 m

**7** There were 24 girls and 36 boys at a sports day. What was the ratio of boys to girls?

**A** 3 to 2 **B** 3 to 5 **C** 2 to 5 **D** 2 to 3

**8** Teresa has $560 that she has put aside to give to her four children. Each child receives $20 per week. How many weeks will the money last?

**A** 7 **B** 8 **C** 9 **D** 10

**9** Water flows into a pool at the rate of 1 litre every 5 seconds. If the water flows for 10 minutes, how much water will flow into the pool in that time?

☐ litres

**10** All of the marbles in a bag were blue or red. The ratio of blue to red marbles was 3 to 5. What fraction of the marbles was blue?

**A** $\frac{3}{5}$ **B** $\frac{2}{5}$ **C** $\frac{5}{8}$ **D** $\frac{3}{8}$

**11** Guido travelled 210 kilometres in three and a half hours. What was his average speed in kilometres per hour?

**A** 35 **B** 60 **C** 70 **D** 84

**12** A car uses fuel at the rate of 7 litres per 100 kilometres travelled. Which is the best estimate for the number of litres needed to travel 420 km?

**A** 30 **B** 40 **C** 50 **D** 60

**13** Adam is paid $28 per hour. If he was paid a total of $1092, how many hours did he work?

☐

**14** Anjuli travels at an average speed of 72 km per hour. How far will she travel in 45 minutes?

☐ km

**15** The ratio of trucks to cars on a freeway was 2 to 7. There were a total of 504 trucks and cars. How many trucks were there?

**A** 72 **B** 112 **C** 252 **D** 392

**16** A car uses petrol at the rate of 8 litres per 100 km travelled. At this rate, how far will the car travel on a full tank of 64 litres?

☐ km

**17** The cost to hire a machine is $40 for the first hour and $30 for every hour (or part of an hour) after that. Sylvia hired a machine at 9:15 am and returned it at 2:50 pm. How much did she have to pay?

$ ☐

**18** Jacob travelled 12 km in 20 minutes. What was his average speed (in kilometres per hour)?

**A** 16 **B** 24 **C** 36 **D** 40

**19** Sacha travelled 105 km at an average speed of 60 km per hour. How long did the journey take?

**A** 1 h 5 min **B** 1 h 25 min

**C** $1\frac{1}{4}$ h **D** $1\frac{3}{4}$ h

**20** Kevin wants to paint his shed. The area to be painted is 48 $m^2$, but the shed needs two coats of paint. If one litre of paint covers 16 square metres, how many litres will Kevin need?

☐ litres



Answers and explanations on pages 110–111

NON-CALCULATOR Intermediate level questions

20 MIN

# Mini Test 19: Mixed Questions

**1** Which of these has half as many faces as edges?
**A** triangular prism
**B** triangular pyramid
**C** rectangular prism
**D** rectangular pyramid

**2** Which fraction is equal to $4\frac{2}{3}$?
**A** $\frac{11}{3}$ **B** $\frac{14}{3}$ **C** $\frac{16}{3}$ **D** $\frac{17}{3}$

**3** Which is the answer to $0.35 \div 0.5$?
**A** 0.7 **B** 0.07 **C** 7 **D** 70

**4** 1 kilolitre and 30 litres is the same as
**A** 100 030 L **B** 130 L
**C** 1.003 kL **D** 1.03 kL

**5** The first four numbers in a pattern are 2, 5, 8 and 11. What is the seventh number in the pattern?

**6** A box holds 5 red, 3 blue and 7 yellow ribbons. Without looking, Shane takes a ribbon from the box. What is the chance that the ribbon is blue?
**A** $\frac{1}{3}$ **B** $\frac{1}{4}$ **C** $\frac{1}{5}$ **D** $\frac{1}{6}$

**7** Boyd has two objects, both made from sticking together four identical cubes.

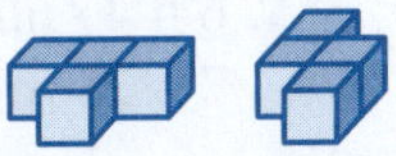

Which shape cannot be formed by joining Boyd's objects?

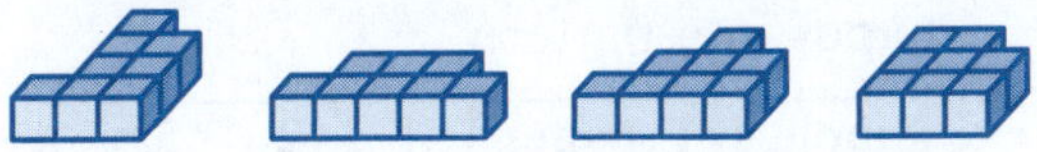

**A** **B** **C** **D**

**8** Jahn has 18 red roses and 6 white roses. What fraction of the roses is white?
**A** $\frac{1}{3}$ **B** $\frac{1}{4}$ **C** $\frac{1}{5}$ **D** $\frac{1}{6}$

**9** This balance shows that 5 blocks have the same mass as 2 discs and 1 block.

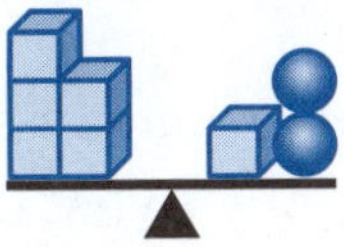

How many blocks balance 3 discs?
**A** 2 **B** 4 **C** 5 **D** 6

**10** A number is multiplied by 7 and then 5 is added. The answer is 47. What was the number?

**11** Which shows an obtuse angle?

**A** **B** **C** **D**

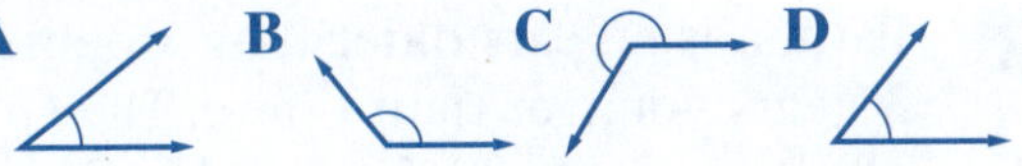

**12** Ellen made this pattern of squares using matches.

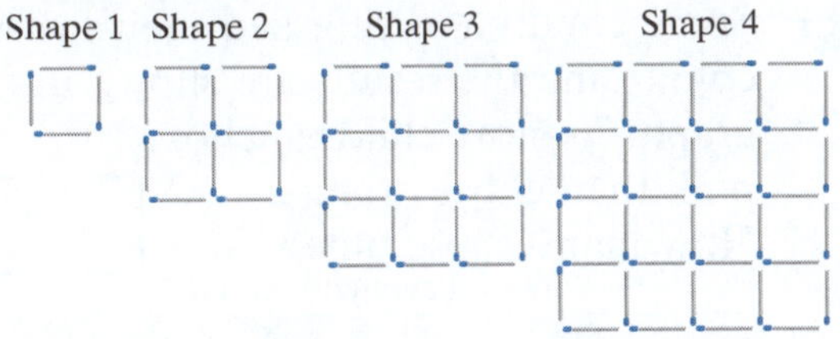


The table shows the number of matches she needed for each part of her pattern.

| Shape | 1 | 2 | 3 | 4 |
|---|---|---|---|---|
| Number of matches | 4 | 12 | 24 | 40 |

How many matches will she need for Shape 5?

**13** Which number is the largest?
**A** 0.2 **B** 0.04 **C** 0.0039 **D** 0.007

**14** What is the perimeter of this shape?
**A** 36 cm **B** 34 cm
**C** 33 cm **D** 31 cm

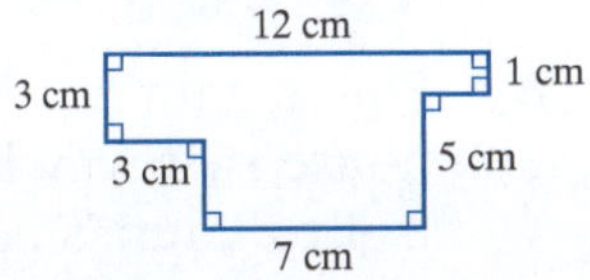


**15** Six men compared their ages. The diagram shows the results.

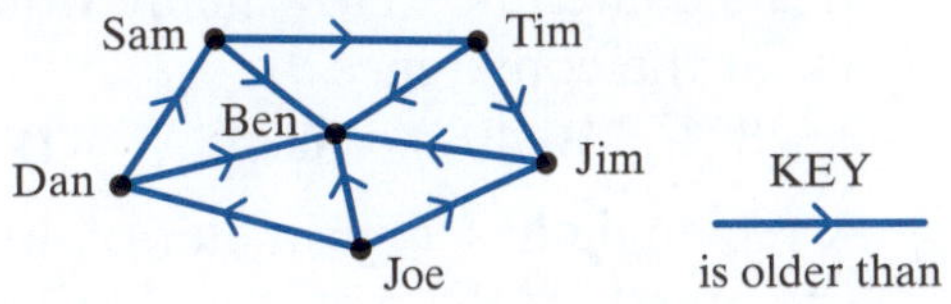


List the men in order from the youngest to the oldest.

**16** Which two clocks show the same time?

# Mini Test 20: Mixed Questions

**1** Ming bought a box of 24 oranges for \$15.60. What was the average price of an orange?

\$ ☐

**2** Fatima is 4 years older than Bonnie but 3 years younger than Claire. The sum of all their ages is 38. How old is Fatima?

**A** 12 **B** 13 **C** 14 **D** 15

**3** Some children were asked their favourite colour and the results are shown in the graph. 20 more children chose pink than chose orange. How many chose blue?

☐

Favourite colour

**4** A car uses 7.5 litres of petrol for every 100 km travelled. How many litres will it use to travel 120 km?

☐ litres

**5** Last year 250 people had a particular disease. This year the number of people with the disease was 125. What percentage decrease is this?

**A** 12.5% **B** 50% **C** 100% **D** 125%

**6** One-eighth of the distance to Laura's house is 16 km. How far is it to Laura's house?

☐ km

**7** The ratio of men to women at a conference is 5 to 3. There are 120 adults at the conference. How many women are at the conference?

**A** 28 **B** 40 **C** 45 **D** 75

**8** A rectangle has a perimeter of 40 cm. The length of the rectangle is 13 cm. What is the width of the rectangle?

☐ cm

**9** The arrow points to a position on the number line.

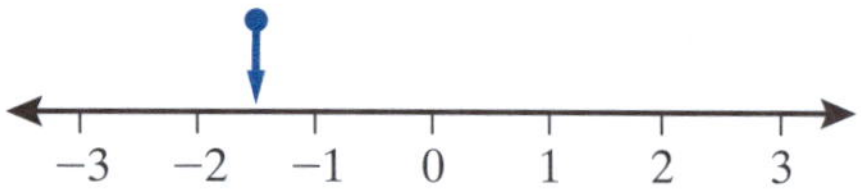


What number is at the position indicated by the arrow?

☐

**10** On a map, 1 cm represents 15 km. Two towns are 5.6 cm apart on the map. What is the actual distance between the two towns?

☐ km

**11** What is the size of the angle?

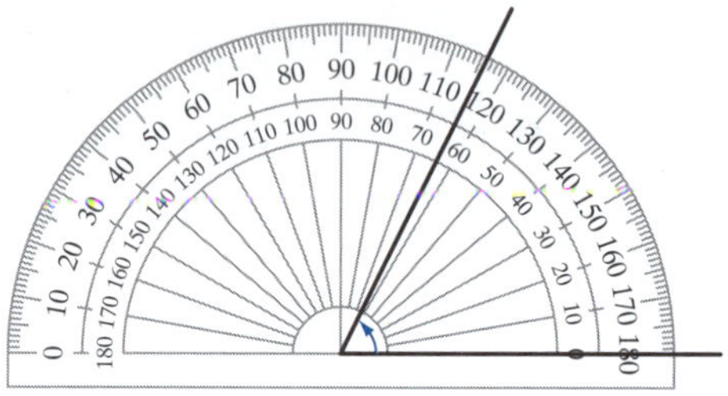

☐ °

**12** A truck can legally carry a maximum load of 8.75 tonnes. How many kilograms can the truck legally carry?

**A** 875 **B** 8075 **C** 8750 **D** 87 500

**13** 6.45 + 5.2 = ☐ + 8.37

What number will make this number sentence true?

☐

**14** How many hours and minutes are between 10.54 am and 6.09 pm on the same day?

**A** 6 h 45 min **B** 7 h 45 min
**C** 7 h 25 min **D** 7 h 15 min

**15** The table shows the number of pins needed to make different numbers of shapes in a pattern.

| Number of shapes | 1 | 2 | 3 | 4 |
|---|---|---|---|---|
| Number of pins | 5 | 8 | 11 | 14 |

How many shapes in the pattern could be made with 32 pins?

☐

**16** Each triangular face of this prism has area 588 cm$^2$.

What is the total area of all the faces?

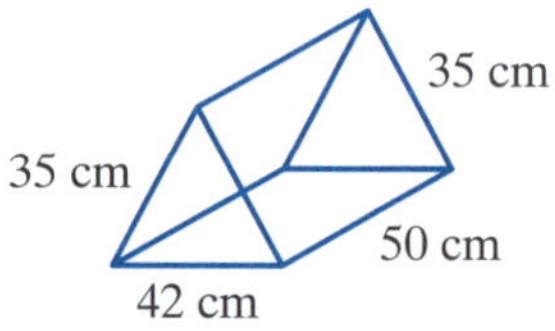


☐ cm$^2$



Answers and explanations on pages 113–114

NON-CALCULATOR **Advanced level questions**

25 MIN

# Mini Test 21: Fractions and Percentages

**1** Which is halfway between $2\frac{1}{5}$ and $2\frac{1}{3}$?

**A** 2 **B** $2\frac{1}{4}$ **C** $2\frac{4}{15}$ **D** $2\frac{1}{2}$

**2** A packet of mixed sweets is made up of jellies, butterscotch and mints. $\frac{2}{5}$ of the sweets are jellies and $\frac{1}{4}$ are mints. What fraction of the sweets is butterscotch?

**A** $\frac{1}{3}$ **B** $\frac{7}{20}$ **C** $\frac{2}{3}$ **D** $\frac{3}{10}$

**3** There are 240 pets at a pet motel. One-quarter of the pets are cats and the rest are dogs. 30% of the dogs are Labradors. How many Labradors are at the pet motel?

**A** 18 **B** 36 **C** 54 **D** 72

**4** Place the fractions $\frac{2}{3}, \frac{3}{4}, \frac{4}{5}, \frac{5}{6}, \frac{7}{10}$ in order from the lowest to the highest.

[ ]

**5** $1\frac{2}{3} \times \frac{4}{5} = ?$

**A** $1\frac{8}{15}$ **B** $2\frac{7}{15}$ **C** $2\frac{1}{5}$ **D** $1\frac{1}{3}$

**6** Which fractions are equal to $\frac{2}{5}$? Select **all** correct answers.

**A** $\frac{4}{25}$ **B** $\frac{12}{30}$ **C** $\frac{16}{40}$ **D** $\frac{28}{80}$

**7** Leo asked 200 people which of four sports they preferred to watch on television. The table shows the results.

| **Sport** | Cricket | Swimming | Netball | Football |
|---|---|---|---|---|
| **Number** | 39 | 51 | 42 | |

What percentage preferred football? [ ] %

**8** Which has the same value as $\frac{13}{4}$?

**A** $3\frac{1}{4}$ **B** $3\frac{3}{4}$ **C** $4\frac{1}{4}$ **D** $4\frac{3}{4}$

**9** 15% of all students received top marks in an exam. $\frac{2}{3}$ of all students who achieved top marks were girls. If 180 students sat for the exam, how many girls got top marks? [ ]

**10** Last week Dave's cows produced 25 000 litres of milk. This week they produced 25 500 litres. What percentage increase is this?

**A** 2% **B** 5% **C** 12.5% **D** 20%

**11** A club has 320 members. 50 of the members are in the club today. The fraction of members who are in the club today is closest to

**A** $\frac{1}{3}$ **B** $\frac{1}{4}$ **C** $\frac{1}{5}$ **D** $\frac{1}{6}$

**12** $\frac{2}{3}$ of a number is 24. What is $\frac{3}{4}$ of the number? [ ]

**13** In 1993, Cameron celebrated his twenty-first birthday. On the same day, his niece Amy was also celebrating her birthday. Amy was $\frac{1}{3}$ of Cameron's age. On their birthday in 2021, what fraction of Cameron's age is Amy?

**A** $\frac{1}{3}$ **B** $\frac{4}{5}$ **C** $\frac{2}{3}$ **D** $\frac{5}{7}$

**14** Which number is the largest?

**A** $\frac{1}{3}$ **B** $\frac{3}{8}$ **C** $\frac{5}{12}$ **D** $\frac{7}{24}$

**15** $15 \div \frac{1}{3} = ?$

**A** 5 **B** 10 **C** 30 **D** 45

**16** A recipe for Anzac biscuits uses $\frac{3}{4}$ of a cup of coconut. This recipe makes three dozen biscuits. Georgia wants to make four dozen biscuits. How many cups of coconut should she use?

**A** 1 **B** $1\frac{1}{4}$ **C** $1\frac{3}{4}$ **D** 3


Answers and explanations on pages 114–116

NON-CALCULATOR **Advanced level questions**

# Mini Test 22: Decimals, Negatives and Ratio

**1** What is the answer to $77 \div 0.7$?

**A** 0.11 **B** 1.1 **C** 11 **D** 110

**2** Riley knows that $18 \times 37 = 666$.
What is $1.8 \times 0.37$?

**A** 66.6 **B** 6.66 **C** 0.666 **D** 0.0666

**3** Write 7%, 0.7, $\frac{1}{7}$ and 0.077 in order from the lowest to the highest.

**4** Which is equivalent to $\frac{13}{100} + \frac{7}{10}$?

**A** 0.137 **B** 0.713 **C** 0.2 **D** 0.83

**5** $1.6 + 3.478 + 0.09 =$

**6** What number is halfway between 0.17 and 0.43?

**A** 0.25 **B** 0.35
**C** 0.3 **D** none of these

**7** $\frac{3 - 0.6}{1 + 0.2} = ?$

**A** 2 **B** 2.2 **C** 2.5 **D** 2.6

**8** What is the seventh number in this pattern? 10, 8.6, 7.2, 5.8, …

**9** The temperature at midnight was −21 °C and at 8 am it was −8 °C.
How much had the temperature risen? °C

**10** At the start of a meeting, the ratio of men to women present was 5 to 4. There were 72 people at the meeting. 5 men and 12 women later left the meeting. What was the ratio of men to women after they had left?

**A** 7 to 4 **B** 5 to 4 **C** 5 to 12 **D** 8 to 5

**11** What is $0.02 \div 0.8$?

**A** 0.025 **B** 0.25 **C** 2.5 **D** 25

**12** Which is **not** equal to 0.6?

**A** $0.15 + 0.45$ **B** $1 - 0.4$
**C** $0.3 \times 0.2$ **D** $0.12 \div 0.2$

**13** Which statements are correct?

**A** $0.2 < 0.05$ **B** $-0.3 > -0.4$
**C** $-0.6 < -0.7$ **D** $-0.18 < -0.153$

**14** The positions of two numbers, (P and Q), are shown on the number line.

What could be the positions of P and Q?

**A** P is −3.4 and Q is −2.6
**B** P is −2.4 and Q is −1.6
**C** P is −3.6 and Q is −2.4
**D** P is −2.6 and Q is −1.4

**15** A business made a profit of \$2500 one month and a loss of \$750 the next month. What was the difference between these results? \$

**16** The ratio of chooks to ducks in Charlie's fowl yard was 8 to 3. Charlie bought 12 more chooks so that he now has 44 chooks. What is the ratio of chooks to ducks now?

**A** 8 to 3 **B** 11 to 3 **C** 20 to 3 **D** 44 to 3

**17** Farid has an old measuring container that measures quantity in gallons. He measured a quantity of fluid and found it to be 2.4 gallons. Farid knows that 1 gallon is equal to 4.54 litres. Which calculation should Farid use to find the quantity of fluid in litres?

**A** $2.4 \div 4.54$ **B** $2.4 \times 4.54$
**C** $4.54 \div 2.4$ **D** none of these

**18** Of the 120 children at a preschool, 75 are boys. What is the ratio of girls to boys?

**A** 8 to 5 **B** 5 to 8 **C** 5 to 3 **D** 3 to 5

**19** The distance from Adamstown to Bennett is 2.7 km and the distance from Adamstown to Dunhill is 7.5 km. Casey is halfway between Bennett and Dunhill.

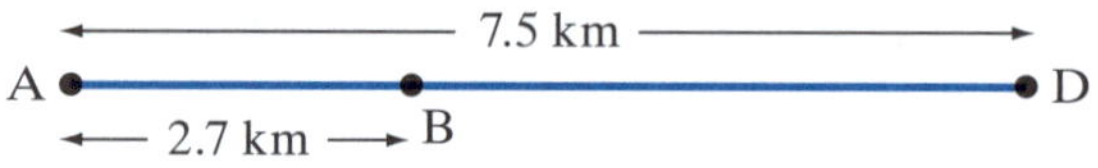


How far is Casey from Adamstown? km

**20** What is the next number in this pattern?

1.2, −1.3, −3.8, −6.3,



Answers and explanations on pages 116–118

NON-CALCULATOR **Advanced level questions**

20 MIN

# Mini Test 23: Patterns and Algebra

**1** 2, 5, 10, 17, 26, ….
What is the next number in this pattern?

**2** Mitchell made this pattern of triangles using matches.

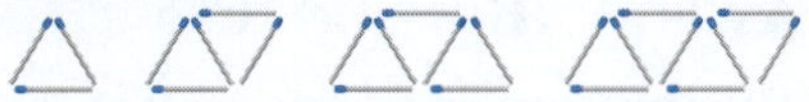

Which rule gives the number of matches needed for each figure?

**A** Multiply the number of triangles by 3.
**B** Multiply the number of triangles by itself and then add 2.
**C** Multiply the number of triangles by 4 and then subtract 1.
**D** Double the number of triangles and then add 1.

**3** 1, 2, 4, 7, 11, …
What is the ninth number in this pattern?

**4** Sofia used the rule 'double the previous number and subtract 3' to work out the numbers in a pattern. The first number in her pattern was 4. Which is **not** a number in the pattern?

**A** 5 **B** 7 **C** 9 **D** 11

**5** A table shows pairs of numbers.

| | | | | | |
|---|---|---|---|---|---|
| **First number** | 1 | 2 | 4 | 7 | 12 |
| **Second number** | 19 | 18 | 16 | 13 | 8 |

What is the second number when the first number is 15?

**A** 3 **B** 4 **C** 5 **D** 6

Use this information to answer questions 6 and 7. The cost per person for a trip is given by the rule: 'Cost = $960 divided by the number of people travelling'.

**6** What is the cost per person when 12 people travel? $

**7** If the cost per person is $30, how many people are travelling?

**8** A pattern has been formed with shapes.

♦♥♠♣♣♠♥♦♦♥♠♣♣♠♥♦♦♥♠…

What is the 50th shape in this pattern?

**A** ♦ **B** ♥ **C** ♠ **D** ♣

**9** 3, 6, 4, 8, 6, 12, 10, 20, 18, …
What is the 12th number in this pattern?

**10** Cooper is drawing a parallelogram *PQRS* on a grid. He has marked the position of *P*, *Q* and *R*. What are the coordinates of *S*?

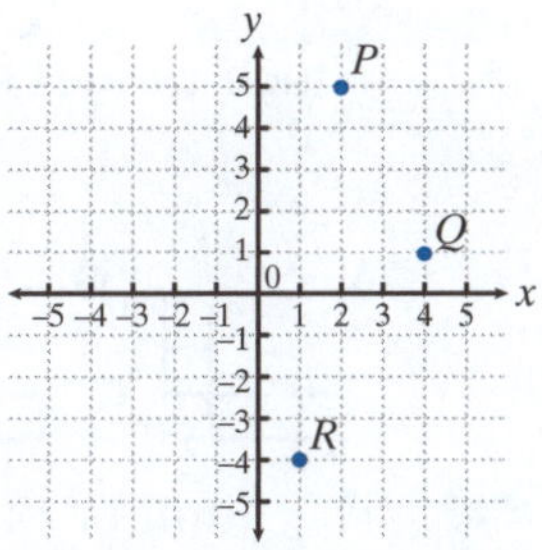


**A** (–1, 0) **B** (0, –1) **C** (–2, 1) **D** (1, –2)

**11** Here are some steps in a pattern:

Step 1: $1 \times 1 = 0 \times 2 + 1$
Step 2: $2 \times 2 = 1 \times 3 + 1$
Step 3: $3 \times 3 = 2 \times 4 + 1$
Step 4: $4 \times 4 = 3 \times 5 + 1$

Using this pattern, what is the answer to the question $99 \times 99$?

**12** Jolene made a pattern of four shapes using matches and then recorded the number of matches needed for each shape in a table:

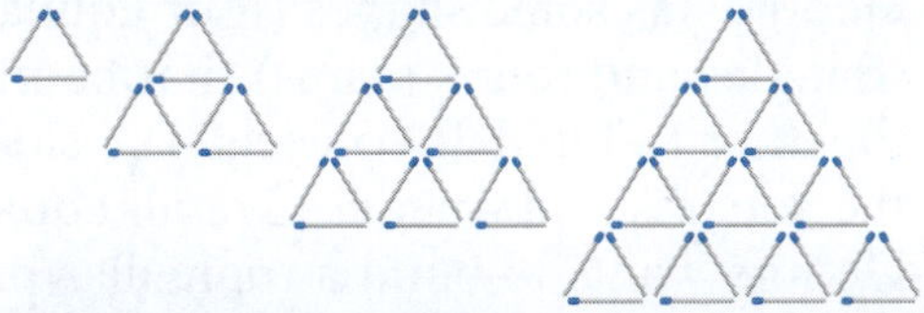

| | | | | |
|---|---|---|---|---|
| **Shape** | 1 | 2 | 3 | 4 |
| **Number of matches** | 3 | 9 | 18 | 30 |

Jolene would need 108 matches to make which number shape?

**13** 2, 5, 8, 11, 14, …
What is the 101st number in this pattern?

**A** 301 **B** 302 **C** 304 **D** 305

**14** Flora used pins to make four shapes in a pattern.

1 2 3 4

How many pins would Flora need for Shape 21 in this pattern?



Answers and explanations on pages 118–120

CALCULATOR ALLOWED **Advanced level questions**

# Mini Test 24: Space

**1** Elizabeth joined some cubes together to make a 3D object. The view from the front and from a side are shown:

Which **cannot** be a top view?

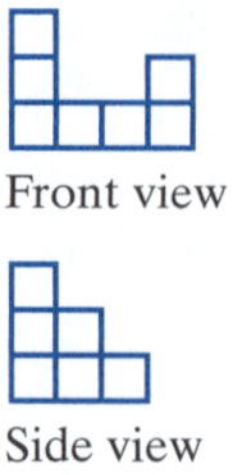


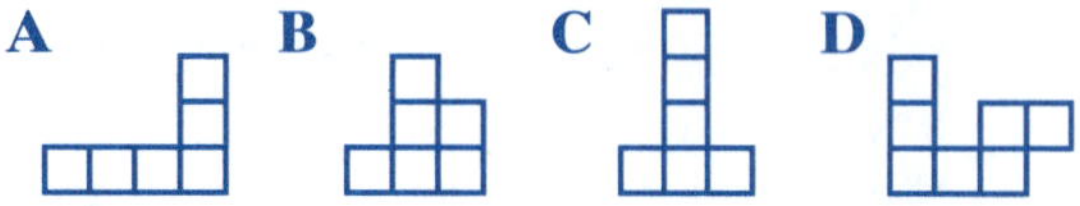


**2** The number of edges of a 3D object is half as many again as the number of faces. Which could be the object?

**A** triangular pyramid
**B** triangular prism
**C** rectangular pyramid
**D** hexagonal prism

**3** Here are two views of the same cube.

Which is the net of the cube?

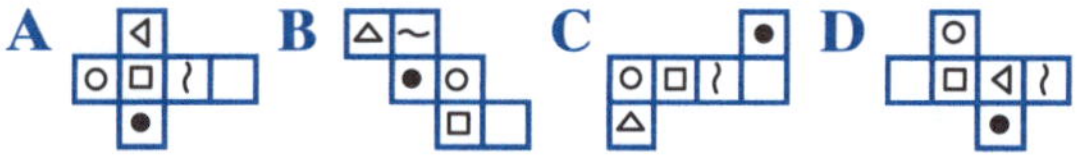


**4** George has some shapes (four equilateral triangles and four squares) that he places together to build 3D objects. The sides of the squares and triangles are all equal. George wants to build a triangular prism, a triangular pyramid and a square-based pyramid. Which statement is correct?

**A** He cannot build a triangular prism.
**B** He cannot build a triangular pyramid.
**C** He cannot build a square-based pyramid.
**D** He can build all three objects.

**5** This prism is made from 60 small cubes that have been joined together. All 6 faces of the prism are painted blue.

The prism is then pulled apart into its 60 individual cubes. Which statements are correct? Select **all** correct answers.

**A** 6 cubes have no blue faces.
**B** 20 cubes have exactly 1 blue face.
**C** 24 cubes have exactly 2 blue faces.
**D** Exactly 6 cubes have 3 blue faces.

**6** The numbers on opposite faces of a cube add to 7.

This cube is rolled over its bottom front edge and then over its (new) bottom right side edge. What face is now on top?

**A** 2 **B** 4 **C** 5 **D** 6

**7** This pattern is formed by rotating the arrow in a clockwise direction.

Through how many degrees is the arrow rotated each time? [ ] °

**8** If a square is cut along one of its axes of symmetry, which of the following is **not** possible as a result?

**A** 2 rectangles
**B** 2 squares
**C** 2 right-angled triangles
**D** 2 isosceles triangles

**9** Mia drew this shape.

Which shape cannot be added to Mia's shape to make a square?

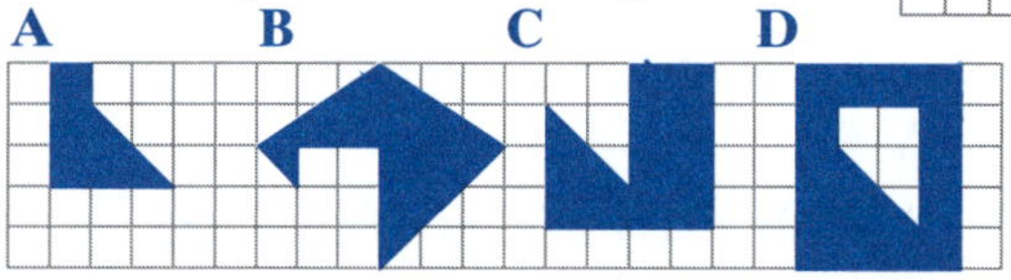


**10** A photo is 12 cm long and 8 cm wide. It is enlarged so that it is now 78 cm long. How wide is it now? [ ] cm

**11** This shape is to be rotated 90° in an anticlockwise direction about the dot and then it is to be reflected in a vertical line. Which diagram shows the result?

**12** Chiara has a cube and two square-based pyramids. All of the square faces are the same size. She sticks the square bases of the pyramids to opposite faces of the cube. How many more edges than faces has the resulting 3D object?

**A** 0 **B** 2 **C** 4 **D** 8



Answers and explanations on pages 120–122

CALCULATOR ALLOWED **Advanced level questions**

25 MIN

# Mini Test 25: Number

**1** The marked price of a laptop is \$850. If this price is reduced by 15%, what price will be paid for the laptop? \$ ______

**2** $\frac{3}{4}$ of all the cows in a paddock had a calf. $\frac{2}{3}$ of all the calves were male. What fraction of all the cows had male calves?

**A** $\frac{1}{2}$ **B** $\frac{8}{9}$ **C** $\frac{5}{12}$ **D** $\frac{1}{12}$

**3** 32% of all the people at a concert were children. If there were 96 children at the concert, how many adults were at the concert? ______

**4** In an opening batting partnership in a cricket match one batsman scored $\frac{3}{5}$ of the runs. The other batsman scored $\frac{1}{3}$ of the runs and the remainder were extras. What fraction of the partnership runs were extras?

**A** $\frac{1}{15}$ **B** $\frac{4}{15}$ **C** $\frac{1}{4}$ **D** $\frac{1}{2}$

**5** A ring was bought for \$2000 and later sold for \$2500. What percentage of the original price was the gain?

**A** 20% **B** 25% **C** 40% **D** 80%

**6** $\frac{3}{5}$ of a number is 45.
What is $\frac{2}{3}$ of the number?

**A** 18 **B** 30 **C** 40 **D** 50

**7** In a netball game, Liz and Kelly played for one team and Sally and Jane played for the other.
Liz scored 13 goals from 20 attempts.
Kelly scored 7 goals from 10 attempts.
Sally scored 9 goals from 15 attempts.
Jane scored 8 goals from 12 attempts.
List the players in order from the one with the highest percentage success rate to the one with the lowest.

______

**8** What is the value of $\frac{18.75 + 12.4}{2.4 - 1.6}$ to the nearest whole number?

**A** 22 **B** 39 **C** 34 **D** 11

**9** Last Friday, 16% of all students at a school were absent. If the school has 900 students, how many of these were **present** last Friday? ______

**10** What number is exactly halfway between 0.6 and 0.04?

**A** 0.05 **B** 0.28 **C** 0.32 **D** 0.5

**11** Over the last six years the value of a painting has increased by 50%. The value of the painting is now \$15 000. What was the value six years ago?

**A** \$7500 **B** \$10000 **C** \$5000 **D** \$12 500

**12** A number was multiplied by itself. The answer was 24 649. What was the number? ______

**13** On Saturday Hayden mowed 60% of the lawn. On Sunday Hayden mowed 30% of the remaining lawn. What percentage of the whole lawn did Hayden mow on Sunday? ______ %

**14** Which does **not** have the same value as $2\frac{3}{4}$?

**A** 2.75 **B** $2\frac{6}{8}$ **C** $\frac{11}{4}$ **D** $2\frac{9}{16}$

**15** A monthly ticket on a train costs \$98.70 and a daily ticket costs \$4.50. What is the greatest number of days that Lily can travel on the train in a month so that it is still cheaper to buy daily tickets rather than a monthly ticket? ______ days

**16** Frieda needed to find the answer to a question but became confused and performed the wrong operations. She was supposed to divide by 3 and then multiply by 0.7 but instead divided by 0.7 and then multiplied by 3. Frieda's answer was 180. What should it have been? ______


Answers and explanations on pages 122–123

CALCULATOR ALLOWED | Advanced level questions | 20 MIN

# Mini Test 26: Patterns and Algebra

**1** 1, 3, 13, 63, 313, ….
What is the eighth number in this pattern?

**2** △ and ♦ stand for numbers. They are related by a rule. What is the rule?

| △ | 2 | 3 | 5 | 8 |
|---|---|---|---|---|
| ♦ | 7 | 16 | 34 | 61 |

**A** ♦ = 6 × △ − 5
**B** ♦ = 9 × △ − 11
**C** ♦ = △ × △ + 3
**D** ♦ = 2 × △ × △ − 1

**3** Jo wants to draw a right-angled triangle. She has placed two points on the grid.
Where could she place her third point? Select **all** correct answers.

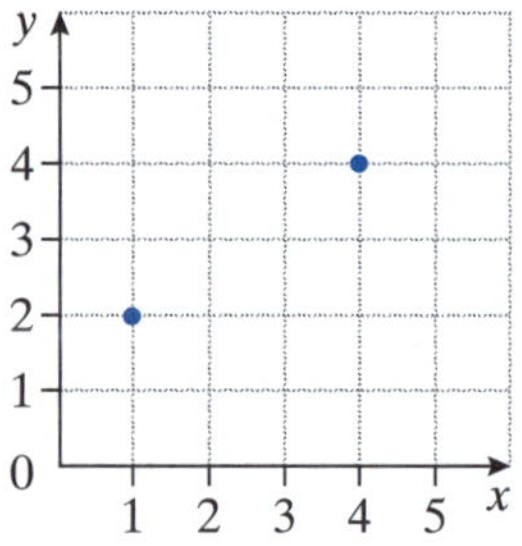


**A** (4, 2) **B** (2, 4) **C** (4, 1) **D** (1, 4)

**4** Dylan used this rule to work out the next number in a pattern: 'Multiply the previous number by 8 and then add 3'. The first 3 numbers in the pattern are 11, 91 and 731. What is the fifth number in the pattern?

Use this information to answer questions 5 and 6.
Ellie made a pattern with pentagons using pins. She drew up a table to show the number of pins needed for different numbers of pentagons.

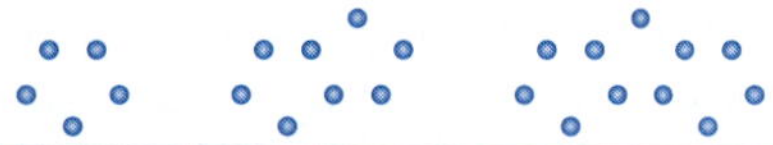

| Number of pentagons | 1 | 2 | 3 |
|---|---|---|---|
| Number of pins | 5 | 8 | 11 |

**5** How many pins are needed for 7 pentagons?

**6** How many pentagons can be made with 60 pins?

**7** What is the next number in this pattern?
1, 2, 5, 26, 677, …

**8** Callie started to draw up a table to show the number of matches needed to make different numbers of squares in a pattern.

| Number of squares | 1 | 2 | 3 | 4 |
|---|---|---|---|---|
| Number of matches | | | 8 | |

Which could be the rule to give the number of matches? Select **all** correct answers.
**A** Double the number of squares and add 2.
**B** Multiply the number of squares by itself and then subtract 1.
**C** Add 5 to the number of squares.
**D** Multiply the number of squares by 4 and subtract 5.

**9** Sergio wrote down the first seven numbers in a pattern: 1, 1, 2, 3, 5, 8, 13
What is the first number in this pattern that is greater than 100?

**10** A rule to calculate the temperature in degrees Celsius (°C) when the temperature in degrees Fahrenheit (°F) is known is:
'First subtract 32, then divide by 1.8'.
An old recipe says to set the oven to 410 °F. What is this temperature in °C? °C

**11** The table shows the **total** number of bricks that can be stacked in different numbers of rows when following a particular pattern.

| Number of rows | 1 | 2 | 3 | 4 | 5 |
|---|---|---|---|---|---|
| Number of bricks | 1 | 3 | 6 | 10 | 15 |

What is the minimum number of rows needed to stack a total of 66 bricks?

**12** The first two numbers in a pattern are 8 and 55. Ben knows that the rule for the pattern is to square the number and then either add or subtract (he can't remember which) a certain number. What is the third number in the pattern?


Answers and explanations on pages 123–125

CALCULATOR ALLOWED   Advanced level questions

# Mini Test 27: Perimeter, Area and Volume

**1** The area of a rectangle is 144 $cm^2$.
The length is 16 cm.
What is the width? ☐ cm

**2** The dimensions of a small rectangular prism are half those of a larger rectangular prism. The volume of the larger prism is 32 cubic metres. What is the volume of the smaller prism in cubic metres?
**A** 16 **B** 8 **C** 4 **D** 2

**3** The area of a square is 121 square centimetres. What is the perimeter? ☐ cm

**4** The perimeter of a rectangle is 28 cm. If the width is 5 cm, what is the area?
**A** 45 $cm^2$ **B** 70 $cm^2$
**C** 90 $cm^2$ **D** 115 $cm^2$

**5** The length of a rectangular prism is 6 m and the width is 5 m. The volume is 90 $m^3$. What is the height? ☐ m

**6** What is the total area of all the faces of this prism?

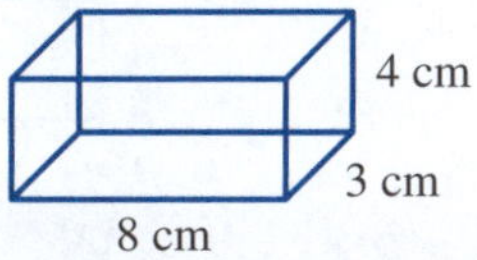


☐ $cm^2$

**7** A rectangle is twice as long as it is wide. Which correctly completes the statement? 'The perimeter is …'
**A** 2 times the width. **B** 6 times the width.
**C** 8 times the width. **C** 12 times the width.

**8** A regular octagon with sides of length 6 cm is made from a piece of wire. If the wire is straightened and then bent to form a square, what is the area of the square? ☐ $cm^2$

**9** Turf is sold in rolls. Each roll is 50 cm wide and, when unrolled, the turf is 3 m long. How many rolls of turf will be needed to cover a rectangular yard that is 20 m long and 12 m wide? ☐

**10** The diagram shows a plan of the floor of a room.

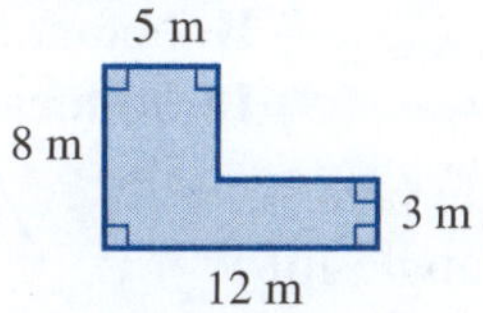


What is the area of the floor? ☐ $m^2$

**11** What is the area of this rectangle?
**A** 240 $cm^2$
**B** 2.4 $m^2$
**C** 24 $m^2$
**D** 240 $m^2$

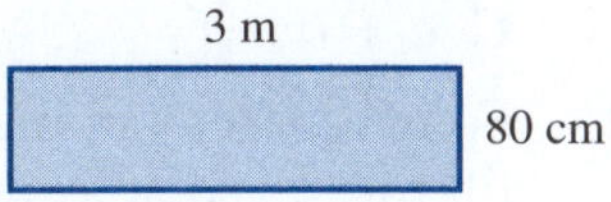


**12** The diagram shows the top view of a swimming pool, which is filled to a depth of 1.2 m.

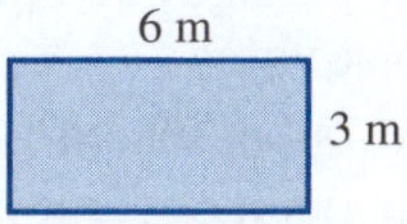


Given that 1 $m^3$ = 1000 L, how many litres of water will the pool hold? ☐ L

**13** The total area of all the faces of a cube is 96 $m^2$.
What is the volume of the cube? ☐ $m^3$

**14** The area of a square is 289 $cm^2$.
What is the perimeter of the square? ☐ cm

**15** Two rectangular prisms have the same volume. The first prism is 5 m long, 4 m wide and 3 m high. What could be the dimensions of the second prism? Select **all** correct answers.
**A** 6 m by 5 m by 2m
**B** 9 m by 3 m by 2 m
**C** 15 m by 2 m by 2 m
**D** 7 m by 3 m by 3 m

**16** The area of a rectangle is 48 $cm^2$.
The width of the rectangle is 6 cm.
What is the perimeter of the rectangle? ☐ cm


Answers and explanations on pages 125–126

CALCULATOR ALLOWED | **Advanced level questions** | 25 MIN

# Mini Test 28: Angles

**1** A polygon has each of its angles equal to 144°. What type of polygon is it?

**A** hexagon **B** octagon
**C** decagon **D** dodecagon

**2** A straight line has divided a triangle into two parts. All of the angles are marked.

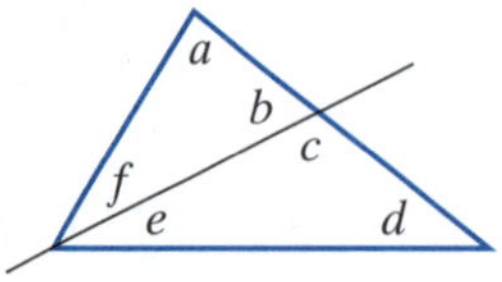


Which expression is **not** equal to 180°?

**A** $a + b + c + d$ **B** $a + d + e + f$
**C** $a + b + f$ **D** $e + c + d$

**3** Four squares can meet and fill the space at a single point. Which of these **cannot** meet at a single point?

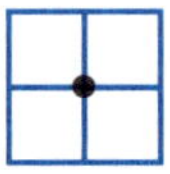

**A** 3 regular hexagons
**B** 1 regular hexagon, 1 equilateral triangle and two squares
**C** 2 regular octagons and 1 square
**D** 1 regular octagon, 2 equilateral triangles and 1 square

**4** What is the value of $x$ in this diagram?

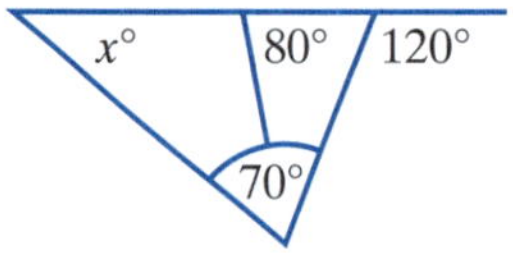


**A** 60 **B** 50
**C** 40 **D** 30

**5** This clock shows that it is 8 o'clock.

What is the size of the **smaller** angle between the two hands?

☐°

**6** A polygon has two angles of 150° and three other equal angles. What size is each of those three remaining angles?

☐°

**7** What is the value of $x$?

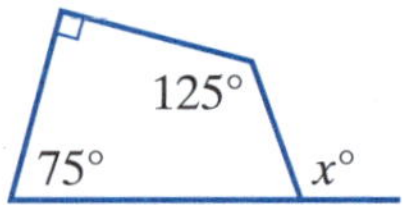


☐

**8** A polygon has two angles of 90° and each of its remaining angles is 120°. What type of polygon is it?

**A** pentagon **B** hexagon
**C** octagon **D** decagon

**9** A quadrilateral is divided into four parts with two straight lines. All of the angles are marked.

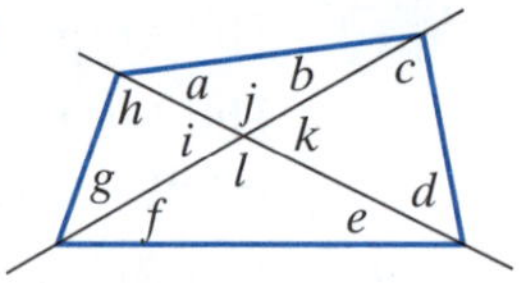


Which expression is equal to 360°?

**A** $a + b + j$ **B** $d + e + f + g$
**C** $i + j + k + l$ **D** $h + g + c + d$

**10** Which description applies to this triangle?

**A** acute-angled, isosceles
**B** acute-angled, scalene
**C** obtuse-angled, isosceles
**D** obtuse-angled, scalene

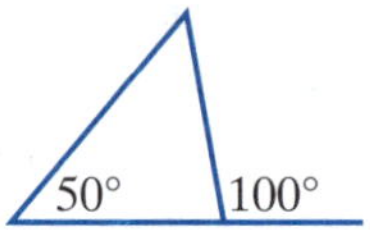


**11** Which could **not** be placed beside the dodecagon to fill the space at P?

**A** a regular hexagon and a square
**B** 2 equilateral triangles and a square
**C** a dodecagon and an equilateral triangle
**D** 4 equilateral triangles

**12** The diagram shows a square and two regular hexagons.

What is the value of $x$?

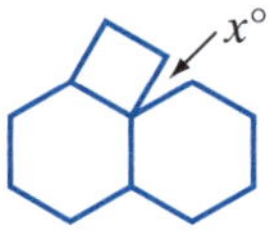


☐

**13** Which is correct?

**A** $x = 25$ and $y = 25$
**B** $x = 50$ and $y = 80$
**C** $x = 50$ and $y = 50$
**D** $x = 80$ and $y = 50$

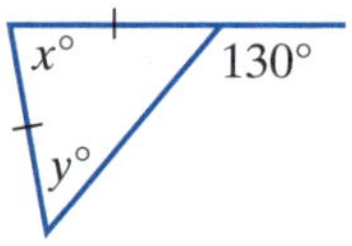


**14** What is the value of $x$?

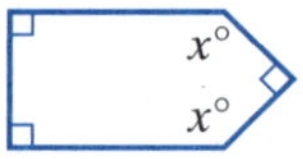


☐

**15** What is the value of $x$?

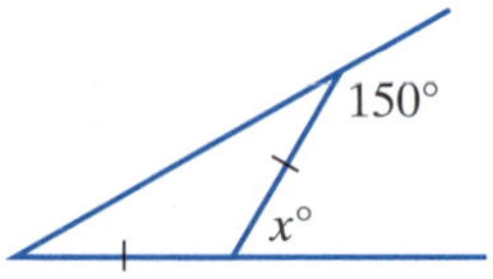


☐

**16** What is the value of $x$?

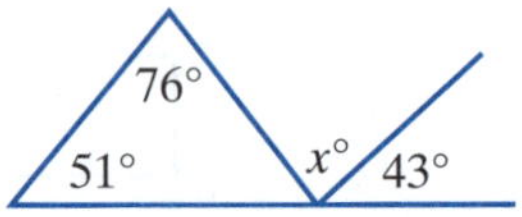


☐



Answers and explanations on pages 126–129

CALCULATOR ALLOWED — Advanced level questions — 20 MIN

# Mini Test 29: Rates and Measurements

**1** Jeremy left home at 8:10 am and drove 240 km at an average speed of 80 kilometres per hour. He then stopped for 50 minutes before setting off again, arriving at his destination at 2:30 pm. The total distance covered was 385 km. What was Jeremy's average speed (in km/h) for the second part of the journey?

**A** 58 **B** 64 **C** 70 **D** 80

**2** The road from Aberdeen to Clovelly passes through Bairnsdale. The distance from Bairnsdale to Clovelly is three times the distance from Aberdeen to Bairnsdale.

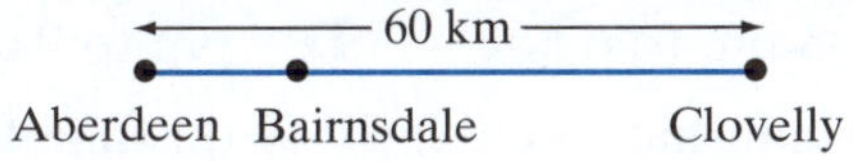


If it is 60 km from Aberdeen to Clovelly, how far is it from Bairnsdale to Clovelly?

**A** 36 km **B** 40 km **C** 45 km **D** 54 km

**3** This bottle has some oil in it. If Ming uses 750 mL of the oil, how much will be left in the bottle? ☐ mL

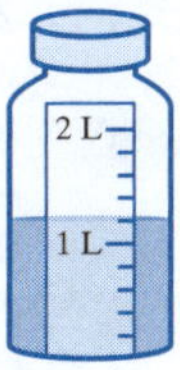


**4** 1 litre of water weighs 1 kg. An empty bottle weighs 20 grams. When full, the bottle weighs 770 grams. How many millilitres of water does the bottle hold? ☐ mL

**5** The diagram shows the route of a 50-km bike race that starts and ends at Putty.

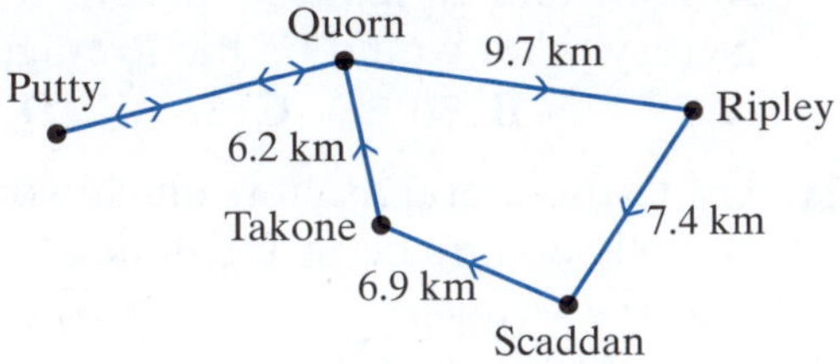


How far is it from Putty to Quorn? ☐ km

**6** A train travelled 594 km at an average speed of 72 km/h. If the journey began at 7:45 am, what time did it finish?

**A** 3:10 pm **B** 4:00 pm
**C** 4:10 pm **D** 5:00 pm

**7** A cricket team needed to score 190 runs from 20 overs. After the first 5 overs the team had scored 49 runs. At what rate (in runs per over) did they need to score the rest of the runs? ☐ runs/over

**8** 1 km and 15 cm is the same as

**A** 1015 cm. **B** 10015 cm.
**C** 100015 cm. **D** 1000015 cm.

**9** Selby left at 8:20 am, travelled 285 km, and arrived at 12:05 pm. What was Selby's average speed for the journey? ☐ km/h

**10** When it is 5:20 am Friday in London it is 3:20 pm Friday in Sydney. When it is 11:45 am Sunday in London it is 6:45 am Sunday in New York. What is the time in New York when it is 10:30 am on Tuesday in Sydney?

**A** 5:30 am Tuesday **B** 7:30 pm Tuesday
**C** 5:30 pm Monday **D** 7:30 pm Monday

**11** Guido ran 42 km at an average speed of 18 km/h. How long did he take?

**A** 2 h 20 min **B** 2 h 33 min
**C** 3 h 40 min **D** 4 h 29 min

**12** Which is the same as 1 hectare and 200 square metres? Give **all** correct answers.

**A** 1200 m$^2$ **B** 10200 m$^2$
**C** 100200 m$^2$ **D** 1.2 ha
**E** 1.02 ha **F** 1.002 ha

**13** Jamie's car uses 9 litres of petrol for every 100 km travelled. He buys $63 worth of petrol at $1.25 per litre. How many kilometres will his car travel on this amount of petrol? ☐ km

**14** A 4-litre container can be filled from a tap in 27 seconds. At that rate, how long will it take to fill a drum that holds 100 litres?

**A** 6 min 45 s **B** 7 min 15 s
**C** 11 min 15 s **D** 11 min 25 s



Answers and explanations on pages 129–130

CALCULATOR ALLOWED | Advanced level questions

# Mini Test 30: Chance and Data

**1** The average (mean) of 7 numbers is 29. One number is left out and the average increases to 32. What number is left out?

**A** 11 **B** 21 **C** 29 **D** 32

**2** Lindsay recorded the colours of cars that went past her house. The results are shown in the table.

| Colour | Number |
|---|---|
| Red | 16 |
| White | 32 |
| Blue | 21 |
| Silver | 12 |
| Other | 15 |

Based on these results, what is the probability that a car is silver?

**A** $\frac{1}{5}$ **B** $\frac{1}{6}$ **C** $\frac{1}{7}$ **D** $\frac{1}{8}$

**3** Each week, Liam's class do a test marked out of 100. Liam's average for the first six tests is 78. What must he score in the seventh test to increase his average to 80?

**4** The eye colour of 200 students was recorded and the results were shown in a sector graph.

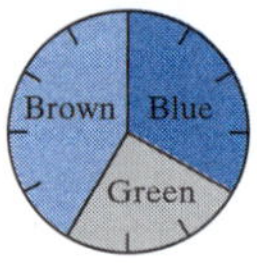


What is the best approximation for the number of students whose eyes are brown?

**A** 50 **B** 65 **C** 80 **D** 95

**5** Corey spins this spinner 60 times.

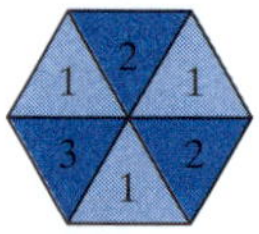


Which table is most likely to show the results?

**A**

| Number spun | Number of spins |
|---|---|
| 1 | 20 |
| 2 | 20 |
| 3 | 20 |

**B**

| Number spun | Number of spins |
|---|---|
| 1 | 20 |
| 2 | 15 |
| 3 | 25 |

**C**

| Number spun | Number of spins |
|---|---|
| 1 | 25 |
| 2 | 20 |
| 3 | 15 |

**D**

| Number spun | Number of spins |
|---|---|
| 1 | 30 |
| 2 | 20 |
| 3 | 10 |

**6** The average (mean) age of 11 cricket players was 32. The 12th man was 20 years old. What was the average age of the 12 players?

**A** 26 **B** 29 **C** 31 **D** 32

**7**

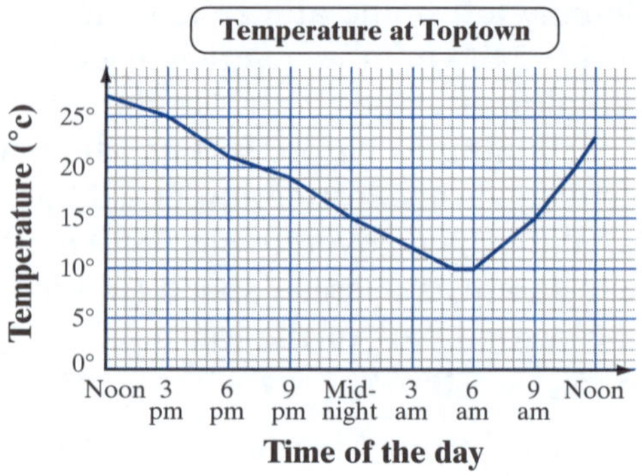


Over which of the time periods below did the temperature **change** the most?

**A** noon to midnight **B** 3 pm to 3 am
**C** 6 pm to 6 am **D** 9 pm to 9 am

**8** A man has been early or on time for work on 15 days and late on 30 days. Based on this information, what is the probability that he will be late?

**A** $\frac{1}{2}$ **B** $\frac{1}{3}$ **C** $\frac{2}{3}$ **D** $\frac{2}{5}$

**9** A few entries in this table are missing.

| | Driver | Passenger | Total |
|---|---|---|---|
| **Men** | 79 | | 106 |
| **Women** | | | |
| **Total** | | 127 | |

How many women were passengers?

**10** The average (mean) age of male workers at a factory is 25 and the average age of female workers is 35. If there were 30 male and 20 female workers at the factory, what would be the average age?

**A** 29 **B** 30 **C** 31 **D** 32

**11** Using these graphs, how much would 100 NZ dollars be in US dollars?

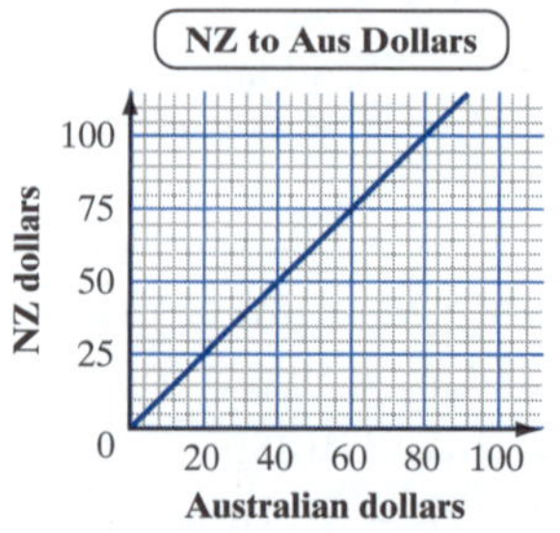


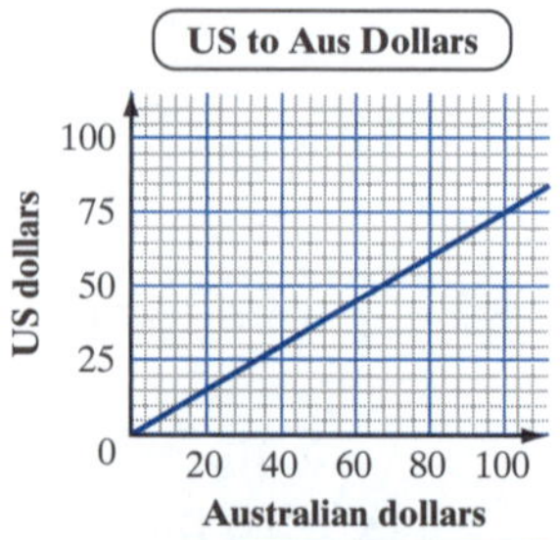


US$



Answers and explanations on pages 130–132

NON-CALCULATOR **Advanced level questions**

25 MIN

# Mini Test 31: Mixed Questions

**1** Which fraction has the same value as $5\frac{3}{4}$?

**A** $\frac{23}{4}$ **B** $\frac{17}{4}$ **C** $\frac{19}{4}$ **D** $\frac{21}{4}$

**2** Which number is exactly halfway between $5\frac{1}{2}$ and 7?

**A** 6 **B** $6\frac{1}{4}$ **C** $6\frac{3}{8}$ **D** $6\frac{3}{4}$

**3** The area of a rectangle is 108 cm$^2$. The width of the rectangle is 9 cm. What is the perimeter of the rectangle? ☐ cm

**4** $(10 + 3) \times 5$ ☐ $(15 - 9) \times (7 + 4)$

Which symbol could be placed in this number sentence to make a true statement? Select **all** correct answers.

**A** $<$ **B** $>$ **C** $=$ **D** $\neq$

**5** Write the fractions $\frac{2}{3}, \frac{3}{4}, \frac{5}{6}$ and $\frac{7}{9}$ in order from lowest to highest.

☐

**6** There were 1500 tickets sold in a competition. 40% of the tickets were blue and the rest were green. One-third of the blue tickets and one-quarter of the green tickets won prizes. How many prizes were won altogether? ☐

**7** Samantha is drawing a rectangle. She has placed three points on the grid. Where should she place the final point?

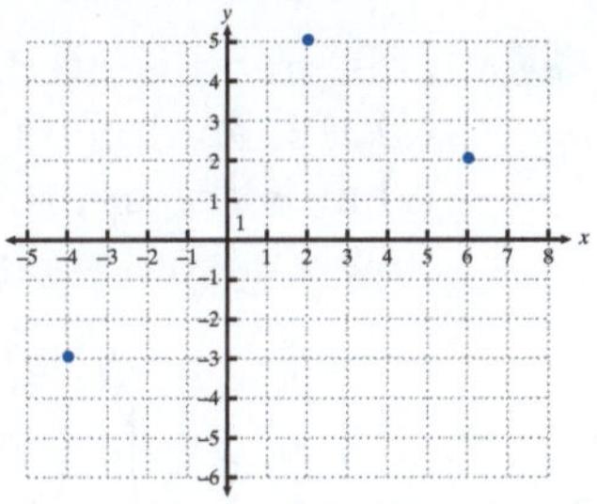

☐

**8** What is the answer to $0.4 \div 0.8$?

**A** 50 **B** 5 **C** 0.5 **D** 0.05

**9** 0.25, 2.5, 4.75, 7, …

What is the eighth number in this pattern? ☐

**10** What is the size of the obtuse angle formed by the hands of this clock? ☐ °

**11** Of the 120 marbles in a jar, $\frac{1}{4}$ are red, $\frac{1}{3}$ are green and the rest are blue. What is the chance that if one marble is taken from the jar without looking it will be blue?

**A** $\frac{5}{7}$ **B** $\frac{2}{7}$ **C** $\frac{7}{12}$ **D** $\frac{5}{12}$

**12** Wendy left at 6:20 am and travelled 245 km, arriving at 9:50 am. What was her average speed? ☐ km/h

**13** Declan used the rule 'double the previous number and add 3' to work out the numbers in his pattern. The first number was 1. What was the fifth number? ☐

**14** A box of chocolates has some dark chocolates and the rest are milk chocolates. The ratio of milk to dark chocolates in the box is 5 to 2. If there are 35 chocolates in the box, how many more milk chocolates are there than dark chocolates?

**A** 10 **B** 15 **C** 20 **D** 25

**15** The temperature at midnight was 4 °C. The temperature at 5 am was 7° colder than it was at midnight. What was the temperature at 5 am? ☐ °C

**16** The volume of a rectangular prism is 160 cubic centimetres. The dimensions of another rectangular prism are half that of the first prism. What is the volume of the second prism?

**A** 20 cm$^3$ **B** 40 cm$^3$ **C** 60 cm$^3$ **D** 80 cm$^3$

**17** Which two of these have the same value?

**A** $1.2 \times 0.01$ **B** $0.12 \div 0.01$

**C** $0.012 \times 100$ **D** $12 \div 1000$



Answers and explanations on pages 132–134

CALCULATOR ALLOWED

Advanced level questions

# Mini Test 32: Mixed Questions

**1** Tamsin received 25 emails yesterday. Today she received 40. What percentage increase is this?

**A** 38% **B** 40% **C** 60% **D** 63%

**2** The ratio of apples to pears in a fruit bin was 8 to 3. There were 192 apples. How many pears were in the bin? ☐

**3** What number is exactly halfway between 37.2 and 50.6? ☐

**4** A polygon has two right angles. Each of its other angles measures 135°. What type of polygon is it?

**A** pentagon **B** hexagon
**C** octagon **D** decagon

**5** Simon left home at 8:40 am and drove until 11:10 am at an average speed of 72 kilometres per hour. If his car uses 8 litres for every 100 km travelled, how much petrol would Simon have used on this trip? ☐ litres

**6** 1 gigalitre is 1000000000 litres.

A lake holds 3.9 gigalitres of water. How many kilolitres is this?

**A** 3900000 **B** 390000000
**C** 39000000 **D** 39000

**7** Jake drew up this table showing the relationship between pairs of numbers.

| **First number** | 1 | 2 | 3 | 4 | 5 |
|---|---|---|---|---|---|
| **Second number** | | 8 | 27 | 64 | 125 |

When the first number is 1, what should the second number be?

**A** 1 **B** 3 **C** 5 **D** 7

**8** The triangle in this diagram is

**A** acute-angled, scalene.
**B** acute-angled, isosceles.
**C** obtuse-angled, scalene.
**D** obtuse-angled, isosceles.

**9** A rule to calculate the distance to the horizon (in kilometres) when the height above sea level (in metres) is known is:

Distance = 5 × square root of (height ÷ 2).

Use this rule to find the distance to the horizon from a point 32 metres above sea level. ☐ km

**10** The value of $\frac{\sqrt{15625}}{5 + 20}$ is

**A** 45 **B** 1 **C** 11.25 **D** 5

**11** The average (mean) age of five children is 8. Their father is 32. What is the average age of the children and their father? ☐

**12** Veronica made a shape using identical cubes joined together. She then drew a top view and a side view.

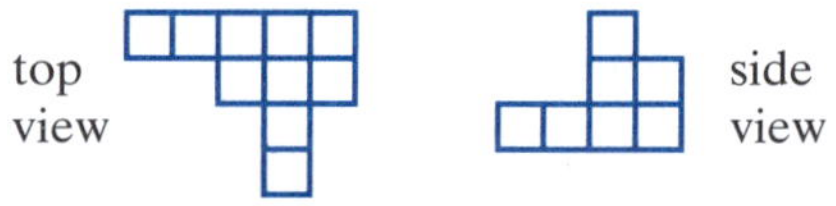


Which could **not** be a front view?

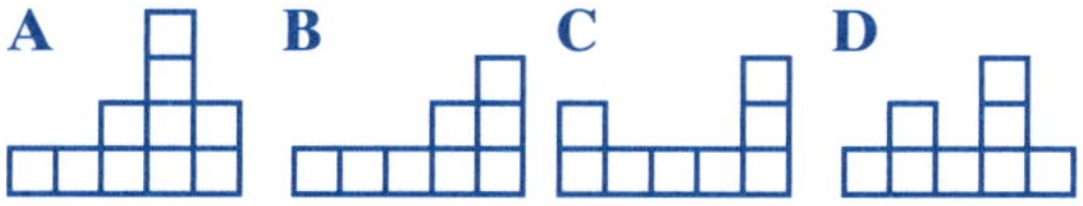


**13** These two rectangles have the same area.

What is the length of the second rectangle? ☐ m

**14** This graph is used to convert amounts in Australian dollars to other currencies. What is the approximate value in New Zealand dollars of 35 British pounds?

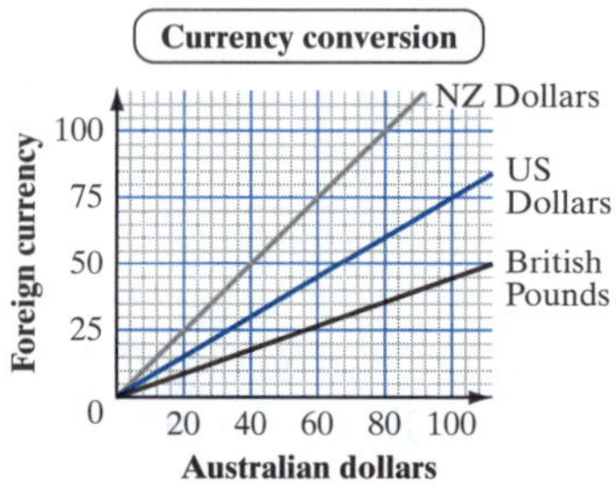


**A** NZ$35 **B** NZ$45 **C** NZ$80 **D** NZ$100




Answers and explanations on pages 134–135

# SAMPLE NAPLAN ONLINE–STYLE NUMERACY TESTS

## DIFFERENT TEST LEVELS

- There are four tests for students to complete in this section. These sample tests have different levels to broadly reflect the sorts of questions students will face in NAPLAN Online, depending on their ability. The following tests are included in this section:
  - one Standard-level Test
  - two Intermediate-level Tests
  - one Advanced-level Test.

## CHECKS

- The NAPLAN Online Numeracy test will be divided into different sections.
- Students will have one last opportunity to check their answers in each section when they have reached the end of that section.
- Once they have moved onto a new section, they will not be able to go back and check their work again.
- We have included reminders for students to check their work at specific points in the Sample Tests so they become familiar with this process before they take the NAPLAN Online tests.

## *Excel Test Zone*

- After students have consolidated their topic knowledge by completing this book, we recommend they practise NAPLAN Online–style questions on our website at www.exceltestzone.com.au.
- Students will be able to gain valuable practice in using an onscreen ruler, protractor and calculator to answer questions.
- Students will also become confident in using a computer or tablet to complete NAPLAN Online–style tests so they will be fully prepared for the actual NAPLAN Online tests.

Year 7 Numeracy

# Sample Online-style Test 1

65 MIN

## Standard level

### Section 1: Non-calculator Questions

A calculator must **not** be used for the eight questions in this part of the test.

---

**Question 1**

Nikita wanted to buy these objects for her study: a lamp for $38.64, a printer for $62.27 and a chair for $83.98.

What is the best way to estimate the total cost?

**A** $30 + $60 + $80

**B** $30 + $70 + $90

**C** $40 + $60 + $80

**D** $40 + $70 + $90

---

**Question 2**

Which one of these does **not** have the same value as $9 \times 4$?

**A** $4 \times 9$ **B** $18 \times 2$ **C** $6 \times 6$ **D** $3 \times 12$ **E** $5 \times 8$

---

**Question 3**

$6 \times \square = 12 \times 7$

What is the missing number?

---

**Question 4**

$32 \times 85 = 2720$

What is $3.2 \times 0.85$?

**A** 2.72 **B** 0.272 **C** 27.2 **D** 0.0272

**Question 5**

What is the value of $6 \times (2 + 3) - (7 - 4)$?

**A** 27 **B** 19 **C** 12 **D** 4 **E** –36

**Question 6**

Which fraction has the same value as $3\frac{2}{5}$?

**A** $\frac{11}{5}$ **B** $\frac{17}{5}$ **C** $\frac{16}{5}$ **D** $\frac{18}{5}$

**Question 7**

What is the answer to $2.4 \div 0.2$?

**A** 12 **B** 1.2 **C** 0.12 **D** 0.012

**Question 8**

The temperature at 6 am was −12 °C.

It was 5 °C warmer at 8 am.

What was the temperature at 8 am?

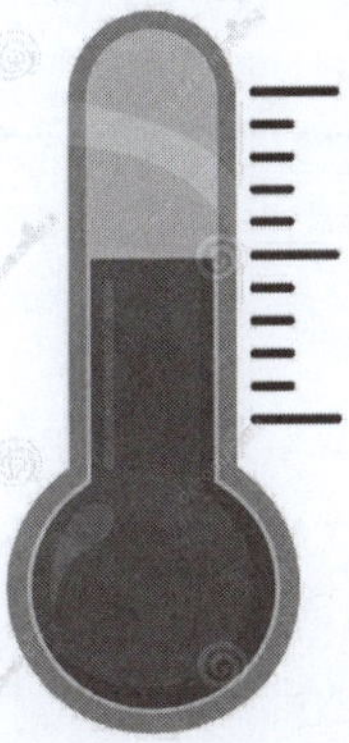

☐ °C

**This is the end of the part where you are not allowed to use a calculator. It would be a good idea to check your answers to the questions in this section before moving on to the other questions.**

## Section 2: Calculator Allowed Questions

### Question 9

Which number is **eight thousand and twelve**?

Circle the correct answer.

**A** 800012 **B** 8000012 **C** 8012 **D** 812000

### Question 10

Which dotted line is a line of symmetry?

### Question 11

What is the value of $\frac{18 \times 12}{24 \div 3}$?

**A** 108 **B** 27 **C** 9 **D** 3

### Question 12

The map shows the position of some of the towns in a local government area.

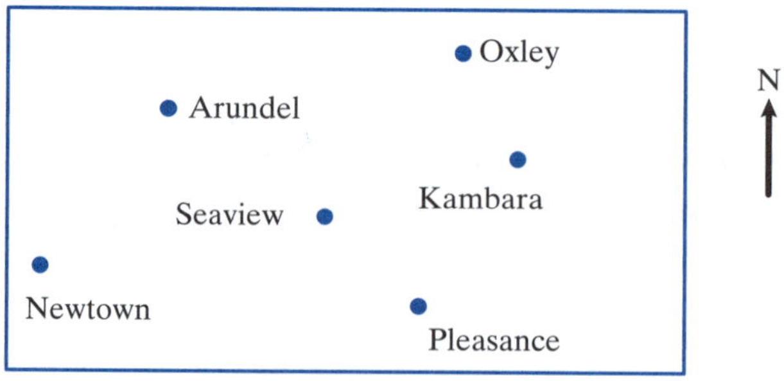


Which town is both North of Kambara and West of Seaview?

**A** Arundel **B** Oxley **C** Newtown **D** Pleasance


Answers and explanations on pages 136–141

**Question 13**

This shape turns about the dot in its centre.

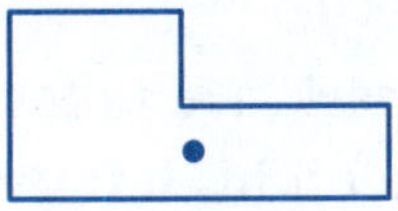

What does it look like after a quarter turn in a clockwise direction?

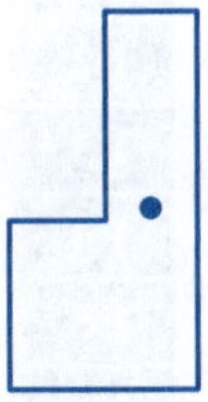

A

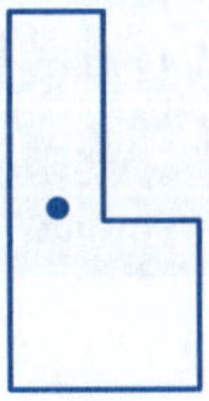

B

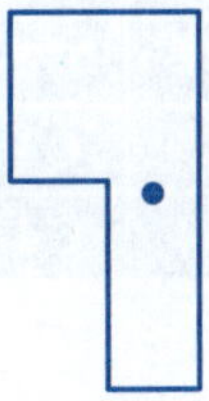

C

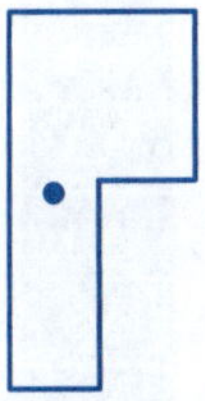

D

**Question 14**

Brandon travelled 24 metres in 3 seconds. What is his average speed in metres per second?

**A** 7 **B** 8 **C** 21 **D** 22

**Question 15**

What fraction of the diagram is shaded?

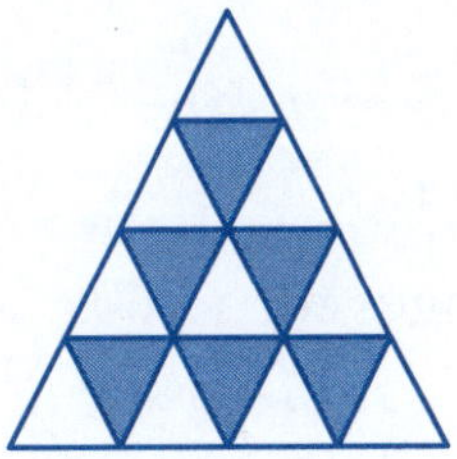

**A** $\frac{1}{2}$ **B** $\frac{3}{5}$ **C** $\frac{6}{13}$ **D** $\frac{3}{8}$

**Question 16**

As a prize on a game show, Zara has to choose one box from six, each of which holds a cash amount.

Which would describe the chance that Zara will choose an amount greater than $1000?

**A** likely **B** unlikely **C** fifty-fifty **D** certain

**It would be a good idea to check your answers to questions 9 to 16 before moving on to the other questions.**

**Question 17**

Rudi drives into the car park. He takes a turn to the right and then turns left. He then parks in the second space on his left.

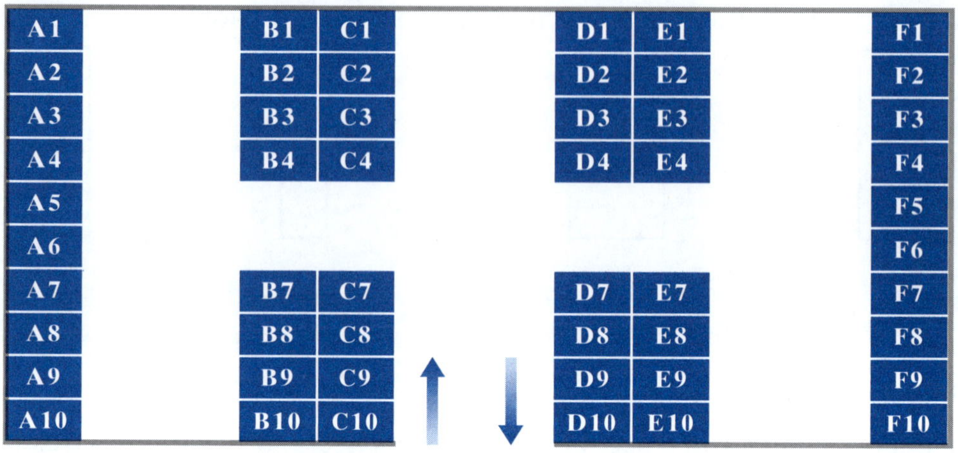


In which space does Rudi park?

**A** A3 **B** B6 **C** E3 **D** F8

**Question 18**

Saffron bought 12 oranges for $6.60.

What was the average price per orange?

$ ______

**Question 19**

A regular octagon is divided into 3 pieces like this:

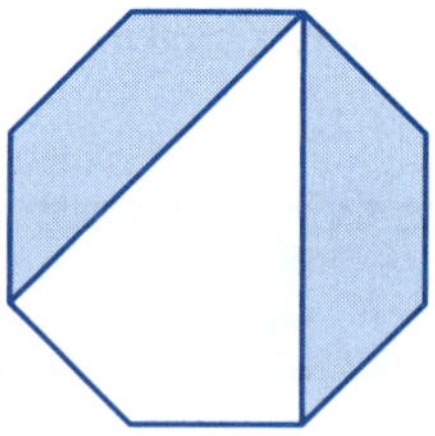

What shape is the unshaded part?

**A** trapezium **B** pentagon **C** hexagon **D** rhombus **E** kite

**Question 20**

Arrange the numbers 3, −8, 12, 0, −17 and −2 in order from lowest to highest.

lowest highest

**Question 21**

What is the value of (18.6 + 2.4) ÷ (1.2 − 0.5)?

**A** 30 **B** 22 **C** 20.1 **D** 16.2

**Question 22**

A piece of pipe needs to be bent to 45°. Which diagram shows the pipe after it is bent?

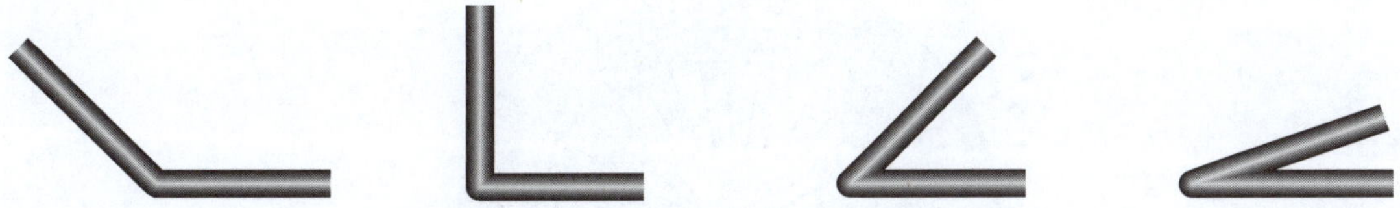

**A** **B** **C** **D**

**Question 23**

Penny drew this shape on a grid. She then halved the height and doubled the width of her shape. What does her shape look like now?

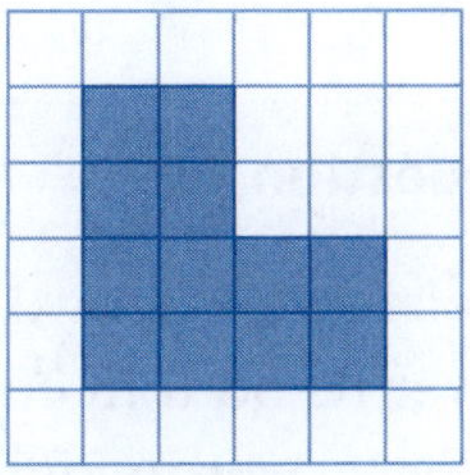

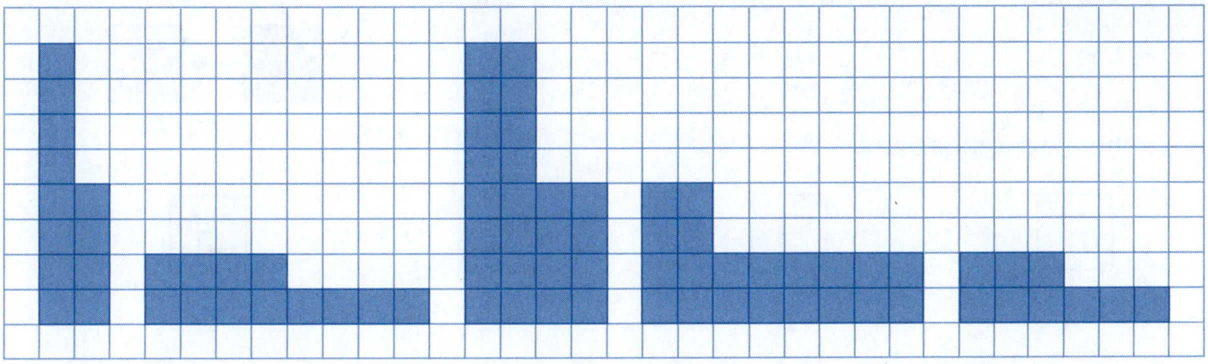

**A** **B** **C** **D** **E**

**Question 24**

Which is the net of a triangular prism? Select **all** correct answers.

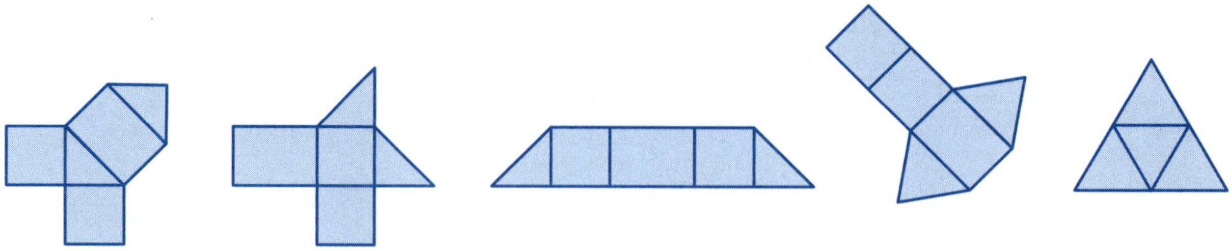

A B C D E

**Question 25**

What is the size of this angle?

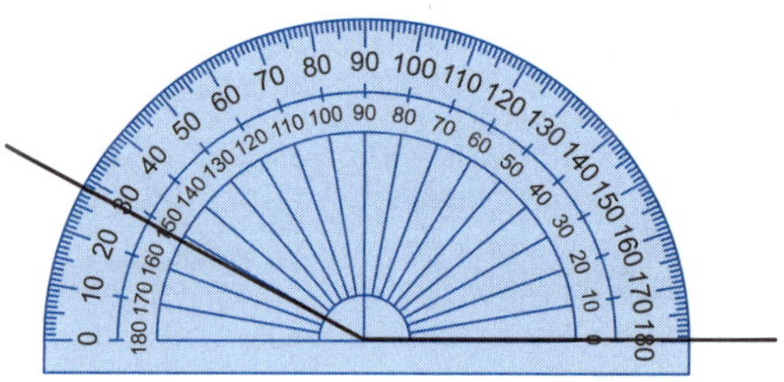

☐ °

**Question 26**

Felicity has a card with her initial printed on it. She flips the card over its bottom edge so that the initial can't be seen and then flips it over its right side so that the initial can be seen again.

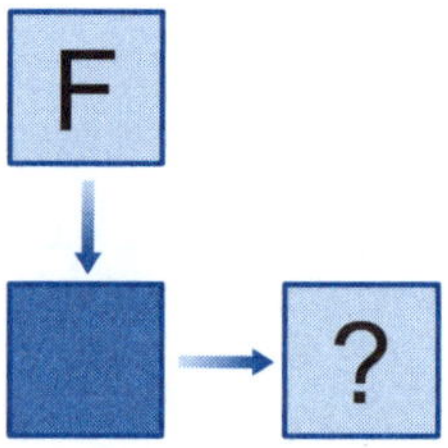


How does the card appear now?

A B C D

**Question 27**

Simone arrived at the station at twenty to ten.

The next train was due to leave at 10:25.

How many minutes did Simone have to wait?

**A** 65 **B** 45 **C** 35 **D** 15 **E** 5

---

**Question 28**

Drew bought a jacket during this sale.
If the marked price of the jacket was $180,
how much did Drew pay?

**A** $27 **B** $83 **C** $153 **D** $165

---

**Question 29**

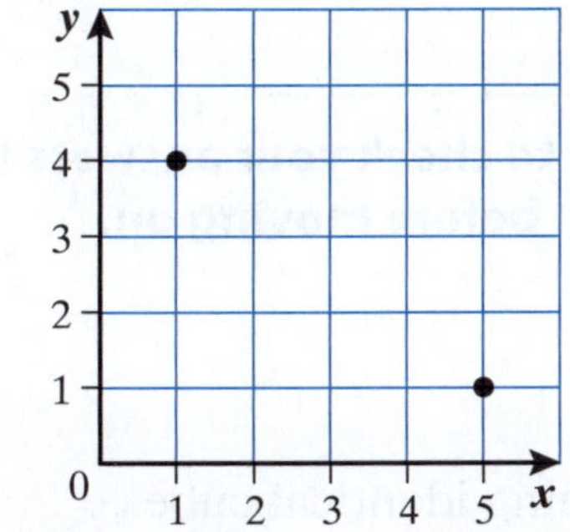


Lincoln is drawing a rectangle on a grid. Two of the corners are shown. He places another corner at (1, 1). Where should the fourth corner go?

**A** (4, 5) **B** (5, 4) **C** (4, 4) **D** (5, 5)

---

**Question 30**

What is the next number in this pattern?

46, 37, 28, 19, ☐


Answers and explanations on pages 136–141

**Question 31**

Anjuli recorded the eye colour of the children in her class in a table.

| Eye colour | Blue | Green | Brown | Other |
|---|---|---|---|---|
| Number | 12 | 5 | 9 | 2 |

She drew a graph to show the information.

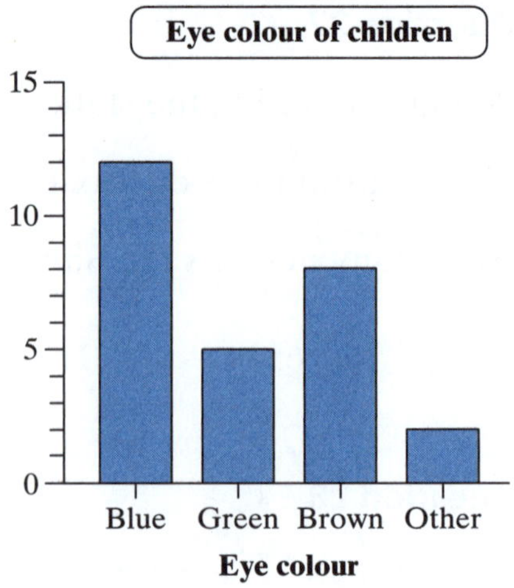


With which eye colour did Anjuli made a mistake in the graph?

**A** Blue **B** Green **C** Brown **D** Other

**Question 32**

In a group of 32 people, 20 are male. What is the ratio of females to males?

**A** 3 to 5 **B** 5 to 8 **C** 3 to 8 **D** 5 to 3 **E** 8 to 5

**It would be a good idea to check your answers to questions 17 to 32 before moving on.**

**Question 33**

Phoebe made this 3D object by joining identical cubes.

What is the view from the left side?

A

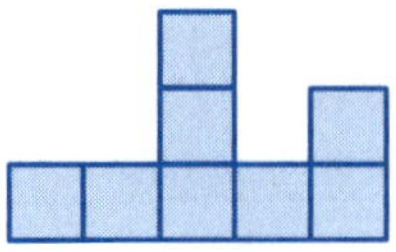

B

C

D

**Question 34**

Which two measurements are the same?

**A** 60 kg **B** 6000000 mg **C** 0.6 t **D** 60000 g

---

**Question 35**

1 m and 30 mm is the same as

**A** 130 m. **B** 130 mm. **C** 1.3 m. **D** 1030 mm.

---

**Question 36**

The number of bicycles sold by a shop in a week is shown in the graph.

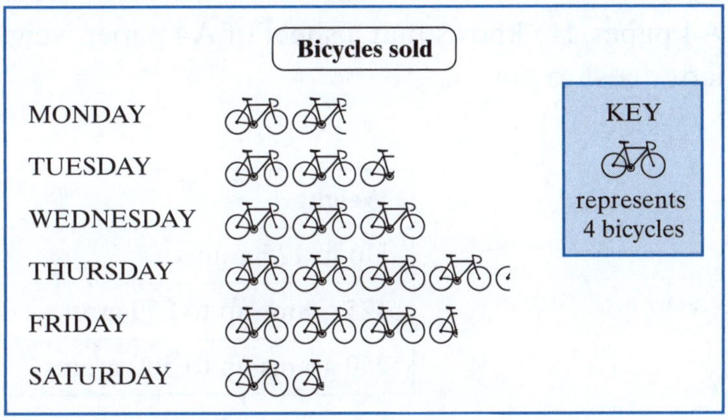


Which statement is correct?

**A** There were 19 bicycles sold on Thursday.

**B** There were 2 more bicycles sold on Friday than on Saturday.

**C** There were 2 fewer bicycles sold on Wednesday than on Tuesday.

**D** There were twice as many bicycles sold on Friday than on Monday.

---

**Question 37**

Which pair of clocks show the same time?

**A** **B** **C** **D**

**Question 38**

□ = 3 and ● = 4

□ + ○ = ● + ●

What number does ○ represent?

**A** 2 **B** 3 **C** 4 **D** 5 **E** 7

**Question 39**

The cost to post a standard letter is $1.00. The cost to post other letters depends on their weight, as shown in the table below. Mitchell wants to send a letter that holds 60 sheets of A4 paper. He knows that a sheet of A4 paper weighs about 5 grams. What should Mitchell's letter cost to post?

| Weight | Cost |
|---|---|
| Up to 125 grams | $2.00 |
| 125 grams up to 250 grams | $3.00 |
| 250 grams up to 500 grams | $5.00 |

What should Mitchell's letter cost to post?

**A** $1.00 **B** $2.00 **C** $3.00 **D** $5.00

**Question 40**

A positive number was multiplied by itself. The answer was 5329.

What was the number? ☐

**Question 41**

What is the value of $x$?

(not to scale)

**A** 35 **B** 40 **C** 55 **D** 70 **E** 75

**Question 42**

Ricky is making this pattern with matches.

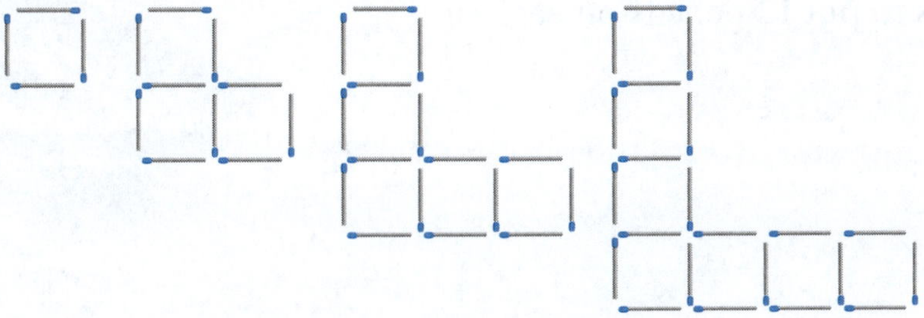

The table shows the number of matches needed for each shape.

| **Shape** | 1 | 2 | 3 | 4 |
|---|---|---|---|---|
| **Number of matches** | 4 | 10 | 16 | 22 |

How many matches will be needed for Shape 5?

**Question 43**

Which of these are smaller than 0.3? Select **all** correct answers.

**A** 0.28 **B** 0.7 **C** 0.0465 **D** 0.009 **E** 0.541

**Question 44**

Jake recorded the number of laps of the oval he ran each afternoon after school.

| **Day** | **Number of laps** |
|---|---|
| Monday | 2 |
| Tuesday | 4 |
| Wednesday | 5 |
| Thursday | 6 |
| Friday | 3 |

What was the mean number of laps per day?

**A** 2 **B** 4 **C** 5 **D** 6 **E** 3

**Question 45**

Lee put 8 pencils on each of 7 desks. He had 4 pencils left over. He wants to put 12 pencils on each of the desks.

How many more pencils does he need?

**Question 46**

This pattern of pentagons has been made with matches.

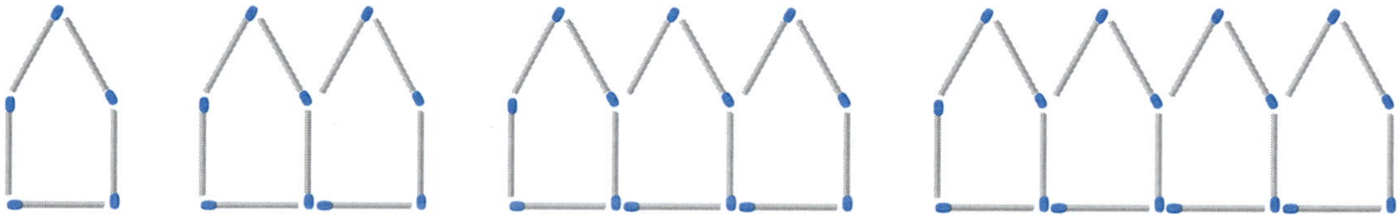

Which is the correct rule for finding the number of matches?

- **A** 5 times the number of pentagons
- **B** 4 times the number of pentagons + 1
- **C** 3 times the number of pentagons + 2
- **D** 2 times the number of pentagons + 3

**Question 47**

Which of these solids have the same number of faces as vertices? Select **all** correct answers.

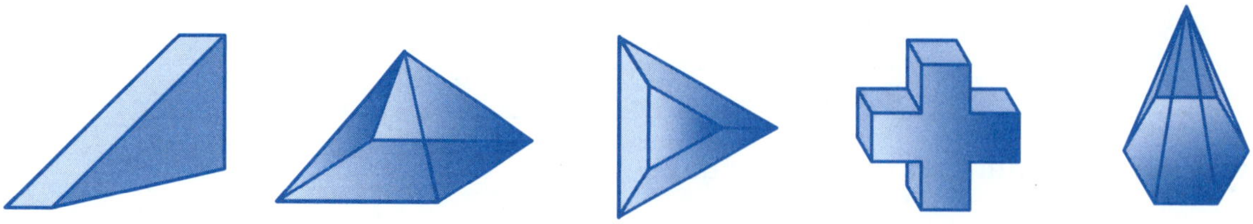

A B C D E


Answers and explanations on page 140

**Question 48**

Write the number of the rectangle, from those below, that matches the given area and perimeter.

| Area 48 $cm^2$<br>Perimeter 28 cm | Area 48 $cm^2$<br>Perimeter 32 cm | Area 45 $cm^2$<br>Perimeter 28 cm | Area 28 $cm^2$<br>Perimeter 32 cm |
|---|---|---|---|
| | | | |

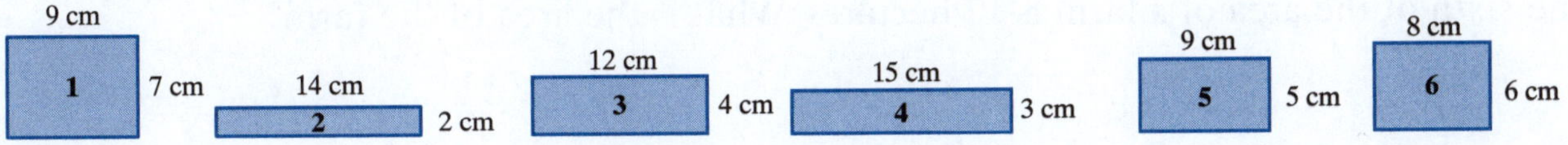


# END OF TEST

# Sample Online-style Test 2

65 MIN

## Intermediate level

### Section 1: Non-calculator Questions

A calculator must **not** be used for the eight questions in this part of the test.

**Question 1**

One-sixth of the area of a farm is 12 hectares. What is the area of the farm?

[ ] ha

**Question 2**

The temperature at 10 pm was −9 °C. At 6 am the next morning it was 12 °C colder.

What was the temperature at 6 am?

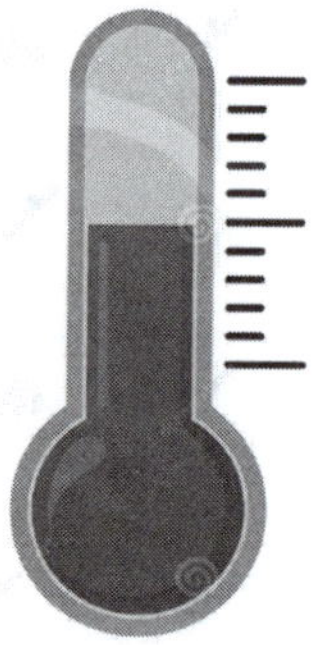

**A** 21 °C **B** 3 °C **C** −21 °C **D** −3 °C

**Question 3**

Sylvie recorded the age of 50 students in this table.

What percentage of the students are 13?

| Age of students | Number of students |
|---|---|
| 11 | 7 |
| 12 | 16 |
| 13 | |
| 14 | 12 |
| Total | 50 |

[ ] %

**Question 4**

What is $2\frac{1}{2} - \frac{3}{4}$?

**Question 5**

$36 \times \square = 16$. What is the value of $\square$?

**A** $\frac{4}{9}$ **B** $\frac{2}{3}$ **C** $\frac{3}{4}$ **D** $\frac{9}{2}$

**Question 6**

Oscar is laying pavers. Each paver is 0.35 m long. There is no gap between the pavers and Oscar wants to cover a distance of 7.7 m.

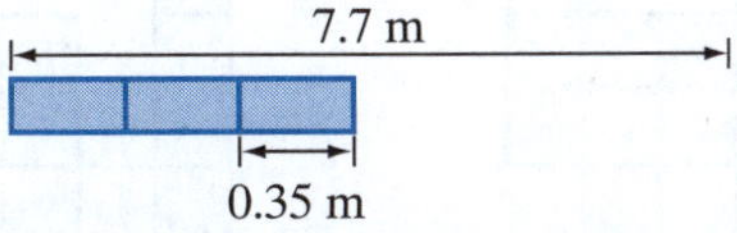


How should Oscar calculate the number of pavers he will need?

**A** $0.35 \times 7.7$ **B** $0.35 \div 7.7$ **C** $7.7 \div 0.35$ **D** $7.7 - 0.35$

**Question 7**

Which has the same value as $4 \times 9$? Select **all** correct answers.

**A** $2 + 10 \times 3$ **B** $5 \times 8 - 4$ **C** $2 \times 12 + 10$ **D** $7 \times 5 + 1$

**Question 8**

Write each fraction as an equivalent fraction in simplest form.

Choose from $\frac{1}{2}, \frac{2}{3}, \frac{3}{4}, \frac{4}{5}, \frac{5}{6}$ or $\frac{7}{12}$.

$\frac{30}{60}$ $\frac{35}{60}$ $\frac{40}{60}$ $\frac{45}{60}$ $\frac{48}{60}$ $\frac{50}{60}$

**This is the end of the part where you are not allowed to use a calculator. It would be a good idea to check your answers to the questions in this section before moving on to the other questions.**

**Section 2: Calculator Allowed Questions**

**Question 9**

Which number is **forty thousand three hundred and two**?

**A** 4302 **B** 40 302 **C** 403 002 **D** 40 000 302 **E** 40 302 000

**Question 10**

Xanthe made a 3D object using 13 identical cubes.

What is the view from the top?

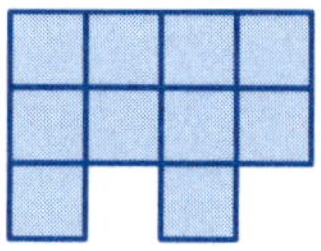

**A**

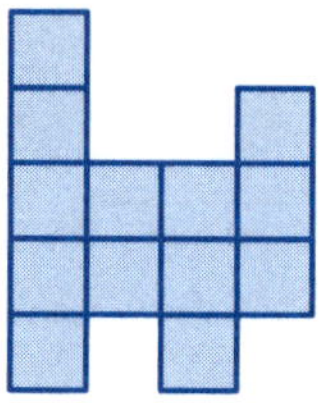

**B**

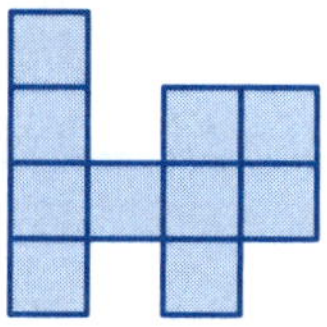

**C**

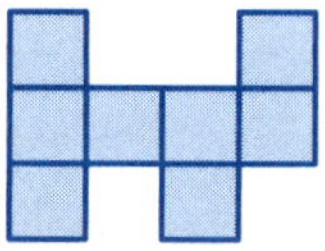

**D**

**Question 11**

Toby drew this shape.

What will it look like after a three-quarter turn in an anticlockwise direction?

**A**

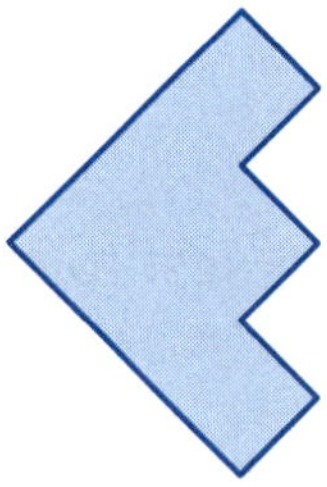

**B**

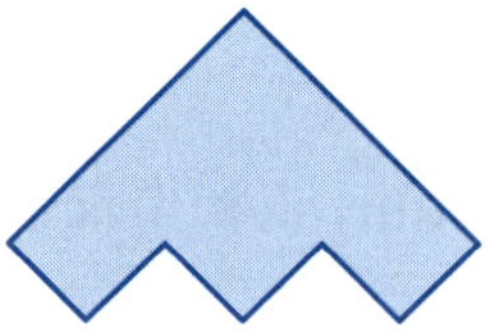

**C**

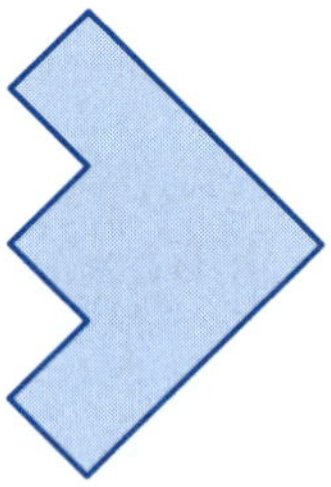

**D**

**Question 12**

What is the value of $\frac{19.72 - 16.8}{0.16 - 0.08}$?

**A** –85.36 **B** 85.36 **C** 190.28 **D** 18.17 **E** 36.5

**Question 13**

Which jug has the most juice?

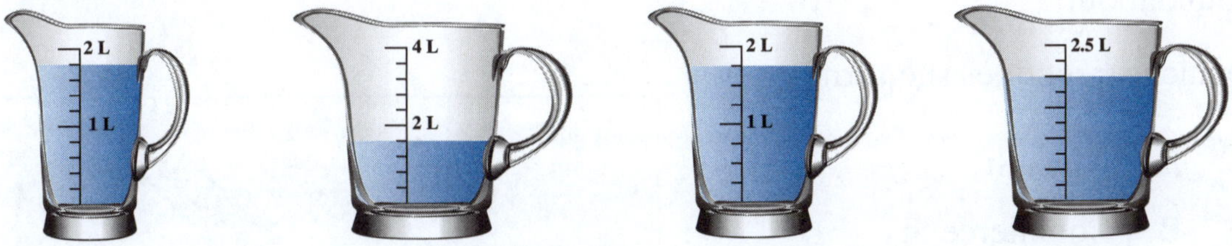


A B C D

**Question 14**

Which arrow is pointing to the location of $-\frac{2}{3}$ on the number line?

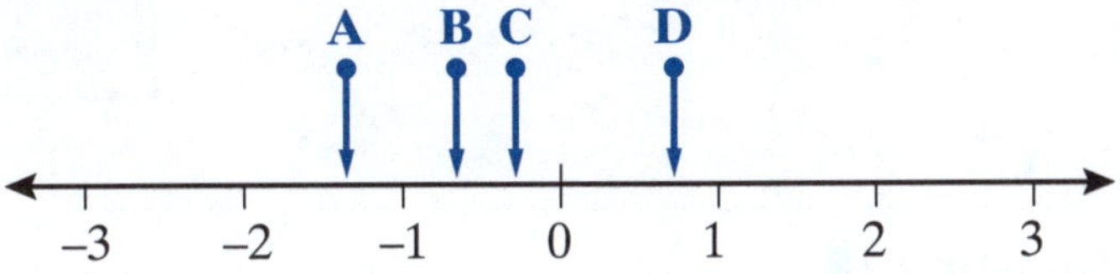


**A** A **B** B **C** C **D** D

**Question 15**

This balance shows that 3 blocks have the same mass as 2 discs.

How many blocks would balance 6 discs and 1 block?

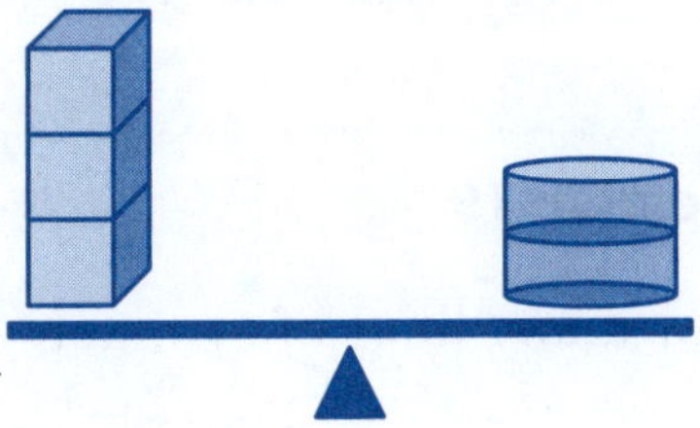

**A** 8 **B** 9 **C** 10 **D** 11

**Question 16**

What is the value of $x$ in this diagram?

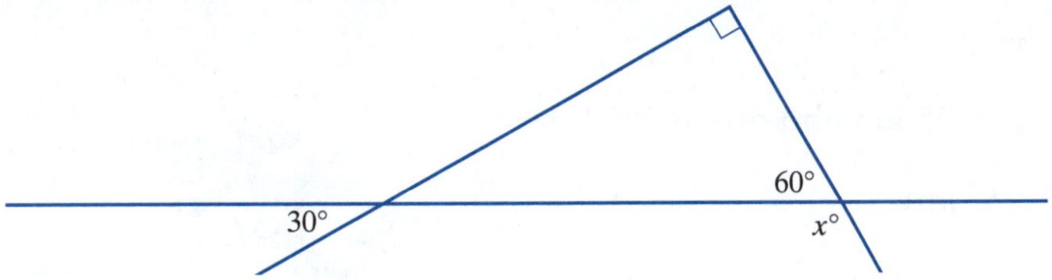


**A** 150 **B** 140 **C** 130 **D** 120

**It would be a good idea to check your answers to questions 9 to 16 before moving on to the other questions.**


Answers and explanations on pages 141–146

**Question 17**

Winona is driving South-East in High Street. She takes the first street on her left after the roundabout.

Into which street does she turn?

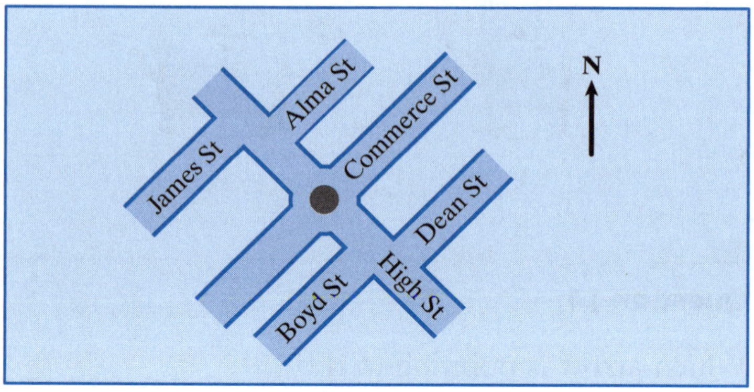


**A** Alma St
**B** Commerce St
**C** Dean St
**D** Boyd St
**E** James St

**Question 18**

1 metre and 15 centimetres is the same as

**A** 1.015 m. **B** 10015 mm. **C** 1150 mm. **D** 1015 cm.

**Question 19**

How many lines of symmetry does this regular octagon have?

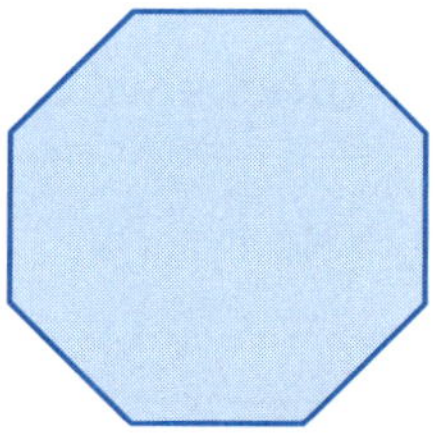

**A** 2 **B** 4 **C** 6 **D** 8 **E** 16

**Question 20**

Tayla left home at 8:35 am and drove 280 km.

She arrived at 12:35 pm.

What was her average speed? ☐ km/h

**Question 21**

This is a map of Jill's farm.

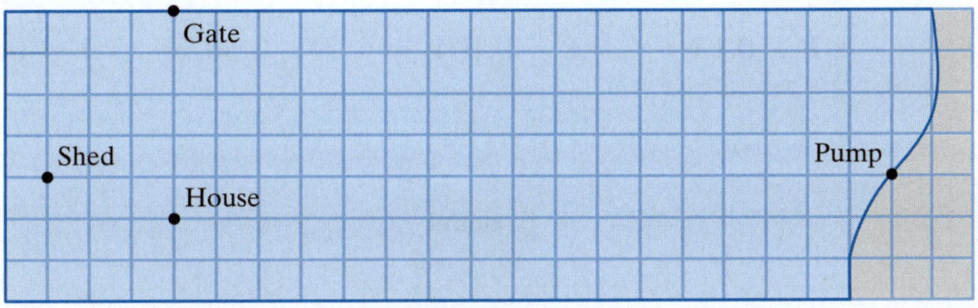


The distance from the gate to the house is 300 metres.

What is the distance from the shed to the pump? [ ] m

**Question 22**

Vincent correctly described the chance of choosing a blue button from a tin of buttons as 'unlikely, but not impossible'. Which could be Vincent's tin?

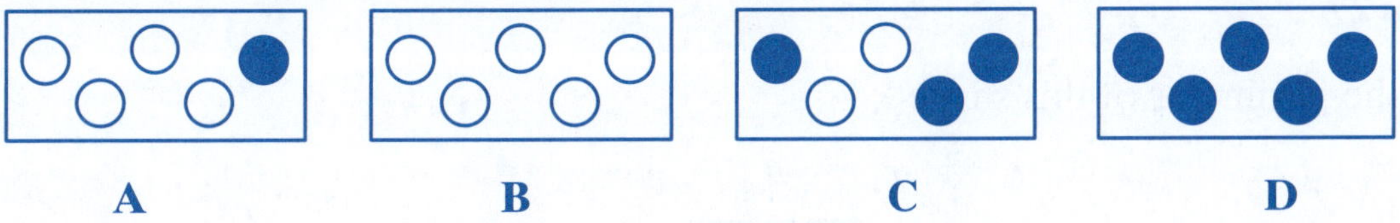


**Question 23**

Amelia thought of a number. She multiplied her number by 4 and then added 3.
Her answer was 47. What was the number that Amelia first thought of?

[ ]

**Question 24**

Which of the marked angles is obtuse? Select **all** correct answers.

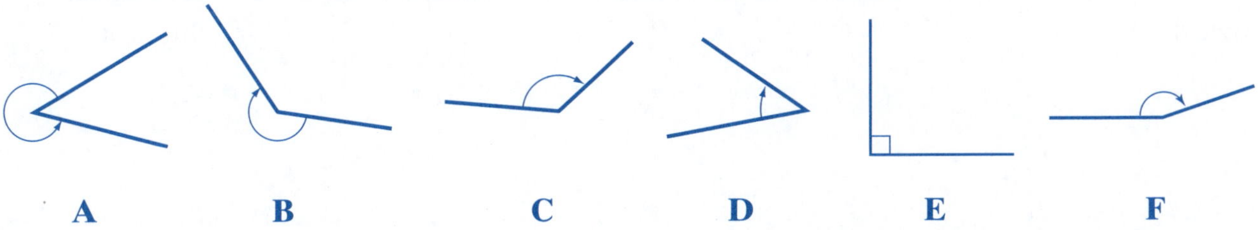

**Question 25**

Which numbers are less than 0.06? Give **all** correct answers.

**A** 0.09 **B** 0.13 **C** 0.104 **D** 0.0078 **E** 0.025

**Question 26**

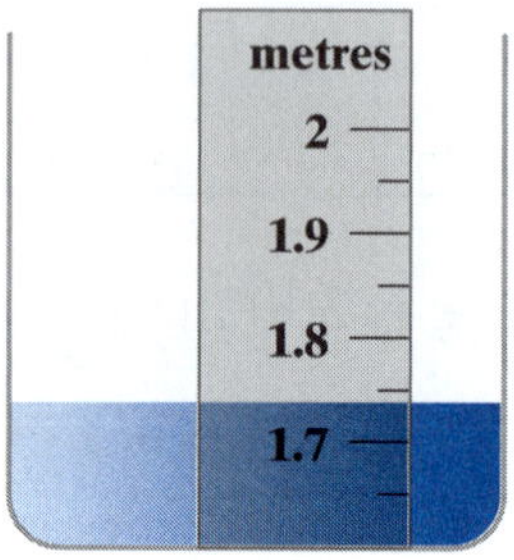


This rule measures the depth of liquid in a vat. How deep is the liquid?

**A** 184 cm **B** 186 cm **C** 176 cm **D** 174 cm

**Question 27**

What is the perimeter of this shape?

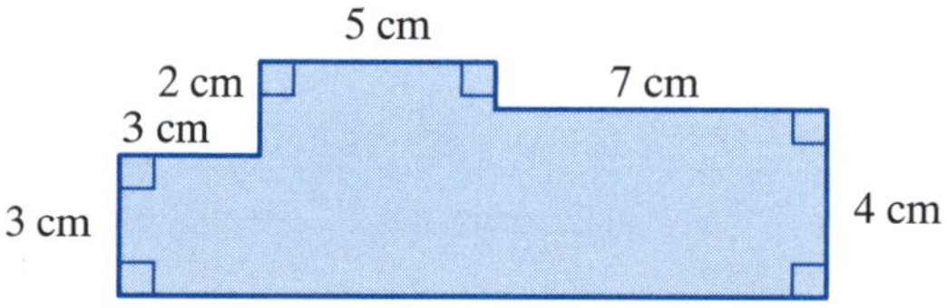


**A** 24 cm **B** 25 cm **C** 39 cm **D** 40 cm **E** 41 cm

**Question 28**

Place the numbers –2.5, −5, 3, 8, −2, 0, −9 and −1.7 in order from lowest to highest.

| | | | | | | | |
|---|---|---|---|---|---|---|---|
| | | | | | | | |

lowest highest

**Question 29**

These 3 rectangular prisms have been drawn on an isometric grid.

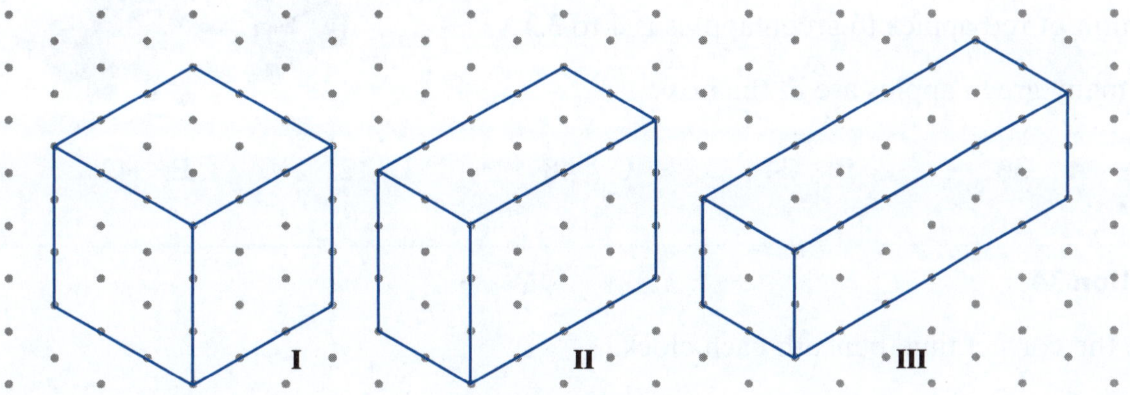


Which have the same volume?

**A** I and II only **B** I and III only **C** II and III only **D** I, II and III

**Question 30**

The perimeter of a square is 36 metres. What is the area of the square?

[ ] $m^2$

**Question 31**

Place the numbers 0.235, 0.3, 0.2178, 0.08, 0.26 in order from highest to lowest.

[ ] [ ] [ ] [ ] [ ]

highest lowest

**Question 32**

A bucket holds 5 yellow, 3 blue and 2 green balls. If one ball is taken from the bucket without looking, what is the chance that it will be green?

**A** $\frac{1}{2}$ **B** $\frac{1}{3}$ **C** $\frac{1}{4}$ **D** $\frac{1}{5}$ **E** $\frac{1}{10}$

**It would be a good idea to check your answers to questions 17 to 32 before moving on.**


Answers and explanations on pages 141–146

**Question 33**

There are 84 apples in a box.

The ratio of red apples to green apples is 4 to 3.

How many green apples are in the box?

**A** 36 **B** 24 **C** 48 **D** 21 **E** 42

**Question 34**

Write the correct time beneath each clock.

Choose from 8:20, 7:20, 3:40, 4:40, 9:15, 3:45, 8:45 or 2:45.

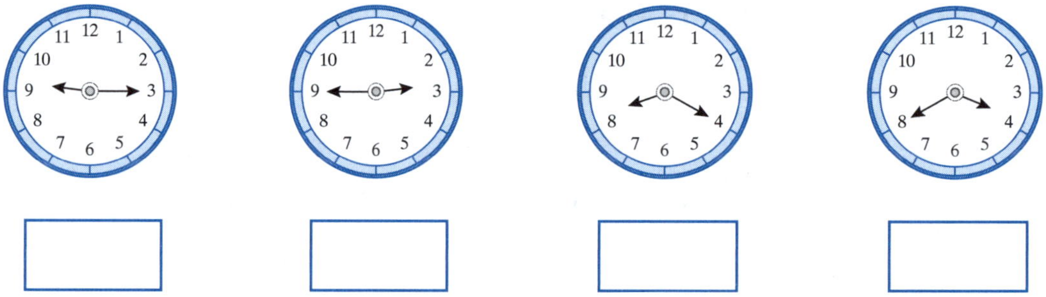

**Question 35**

A photo is 12 cm long and 8 cm wide. It is enlarged so that it is now 48 cm wide.

How long is the enlarged photo?

[ ] cm

**Question 36**

Which is the same as 2 kilolitres and 50 litres? Select **all** correct options.

**A** 200050 L **B** 2050 L **C** 250 L **D** 2.5 kL **E** 2.05 kL

**Question 37**

Rita recorded the colour of 100 cars. This graph shows the results.

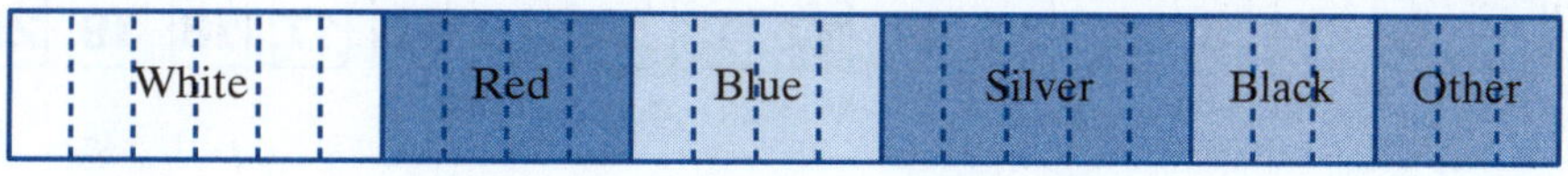


How many more white cars than red cars were recorded?

[ ]

**Question 38**

3 hours and 15 minutes is the same as

**A** 3.15 h. **B** 3.25 h. **C** 315 min. **D** 325 min.

**Question 39**

The square bases of two identical pyramids are joined together. Compared to **one** of the original pyramids, the new object has

**A** 3 more faces and 4 more edges.
**B** 3 more faces and 8 more edges.
**C** 4 more faces and 4 more edges.
**D** 4 more faces and 6 more edges.
**E** 5 more faces and 8 more edges.

**Question 40**

The average (mean) of six numbers is 4.

When one of the numbers is left out the average changes to 3.

What number is left out?

**A** 3 **B** 4 **C** 6 **D** 7 **E** 9


Answers and explanations on pages 141–146

**Question 41**

Circle all the cards that show prime numbers.

| 1 | 2 | 3 | 4 | 5 | 6 | 7 | 8 | 9 | 10 | 11 | 12 | 13 | 14 | 15 | 16 | 17 | 18 | 19 | 20 |
|---|---|---|---|---|---|---|---|---|---|---|---|---|---|---|---|---|---|---|---|

**Question 42**

Saxon had an average (mean) mark of 79 for his first 3 tests. In his fourth test he scored 87. What was his new average?

**A** 81 **B** 82 **C** 83 **D** 84

**Question 43**

A pattern is formed using the rule: 'Halve the number and then subtract 1'. The first number in the pattern is 78. What is the fifth number?

**A** 3 **B** 5 **C** 8 **D** 18

**Question 44**

$X$ is the point (2, 2). Write the coordinates for the points $P$, $Q$, $R$ and $S$. Choose from the following: (0, 1), (1, 0), (−1, 3), (−3, 1), (−1, −3), (−3, −1), (3, −3) and (−3, 3).

$P$ [ ] $Q$ [ ]

$R$ [ ] $S$ [ ]

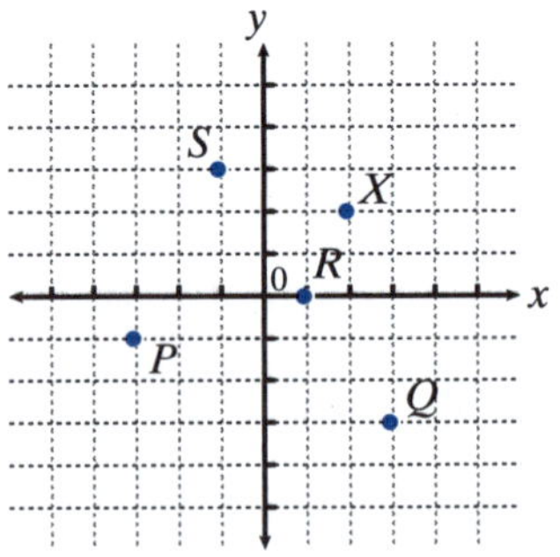



Answers and explanations on pages 141–146

**Question 45**

On a racetrack, Leanne drove at an average speed of 180 kilometres per hour.

If she drove for five minutes, what distance did she travel?

[ ] km

**Question 46**

There are 700 students at a senior college. 60% of the students are boys. 40% of the boys are in Year 12. How many boys are in Year 12 at the college?

[ ]

**Question 47**

A car uses petrol at the rate of 8 litres per 100 km travelled.

How many litres of petrol will the car use to travel 750 km?

**A** 45 **B** 60 **C** 78 **D** 94 **E** 107

**Question 48**

Write a fraction in each box to complete a pattern where the rule is to increase by the same amount each time.

Choose from $\frac{1}{2}, \frac{1}{3}, \frac{1}{4}, \frac{1}{5}, \frac{1}{6}$ or $\frac{1}{8}$.

| | | | | | |
|---|---|---|---|---|---|
| $\frac{1}{12}$ | | | | $\frac{5}{12}$ | |

**END OF TEST**

# Year 7 Numeracy

# Sample Online-style Test 3

65 MIN

## Intermediate level

### Section 1: Non-calculator Questions

A calculator must **not** be used for the eight questions in this part of the test.

**Question 1**

A mattress had a marked price of $800. Vera bought the mattress in a sale where the shop gave 20% off the marked price. How much did she pay?

**A** $600 **B** $640 **C** $720 **D** $780

**Question 2**

250 children were asked whether or not they walked to school.

The table shows some of the results.

| | Walked | Did not walk |
|---|---|---|
| **Boys** | 73 | 52 |
| **Girls** | | 64 |

How many girls walked to school?

**Question 3**

One-fifth of the mass of a bale of hay is 120 kg.

What is the mass of the full bale?

kg

**Question 4**

What is the value of $12 + 8 \div 4 - 2 \times 3$?

**A** 36 **B** 1 **C** −1 **D** 8 **E** 9

**Question 5**

Rema arranges some chairs for a meeting.

If there are seven rows of chairs in this pattern, how many chairs are there altogether?

**A** 49 **B** 56 **C** 63 **D** 70

---

**Question 6**

4.8, 4.75, 4.7, 4.65, …

What is the rule for this pattern?

**A** increase by 0.5
**B** increase by 0.05
**C** decrease by 0.5
**D** decrease by 0.05

---

**Question 7**

Which of these number sentences are equal to 1? Select **all** correct answers.

**A** $3 - 4 \times 2 + 6$
**B** $12 \div 6 \times 2$
**C** $(17 + 18) \div (7 \times 5)$
**D** $4 \times (21 - 12) \div 6^2$
**E** $-8 + 2 - 5 + 6 + 4$

---

**Question 8**

Write the simple fraction that is equivalent to each percentage.

Choose from $\frac{2}{3}, \frac{3}{4}, \frac{4}{5}, \frac{7}{10}, \frac{13}{20}$ or $\frac{18}{25}$.

| 65% | $66\frac{2}{3}$% | 70% | 72% | 75% | 80% |
|---|---|---|---|---|---|
| | | | | | |

---

**This is the end of the part where you are not allowed to use a calculator.**
**It would be a good idea to check your answers to the questions in this section before moving on to the other questions.**

## Section 2: Calculator Allowed Questions

### Question 9

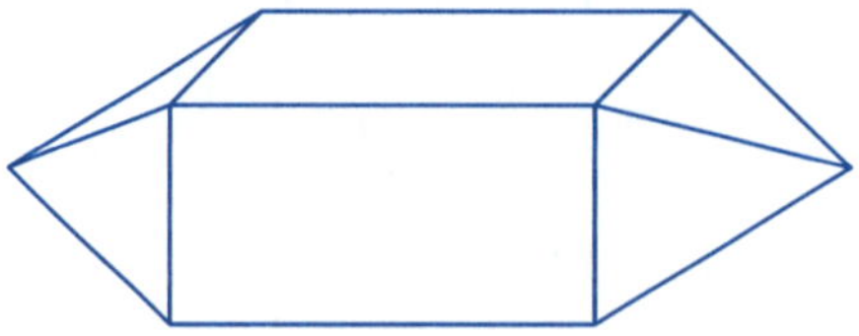

Quentin has made an object by sticking 2 rectangular pyramids to the ends of a rectangular prism. How many edges has Quentin's object?

**A** 12 **B** 16 **C** 20 **D** 28

### Question 10

The temperature in a refrigerator was 3 °C and in a freezer −19 °C.

What is the difference between these temperatures? ☐ °C

### Question 11

Travis was driving along Smith Street towards Dennis Drive.

In what direction was he travelling?

**A** North-East **B** North-West **C** South-East **D** South-West

### Question 12

The population of a country is 3 582 517. What is this population to the nearest thousand?

**A** 4 000 000 **B** 3 500 000 **C** 3 600 000 **D** 3 582 000 **E** 3 583 000


Answers and explanations on pages 146–151

**Question 13**

Eli used some of these cards to make the number three hundred thousand and twenty-eight.

| 0 | 0 | 0 | 0 | 0 | 2 | 2 | 2 | 3 | 3 | 8 |
|---|---|---|---|---|---|---|---|---|---|---|

What number should Eli have made?

---

**Question 14**

Which shows an angle of about 120°?

A B C D

---

**Question 15**

The arrow points to a position on the number line.

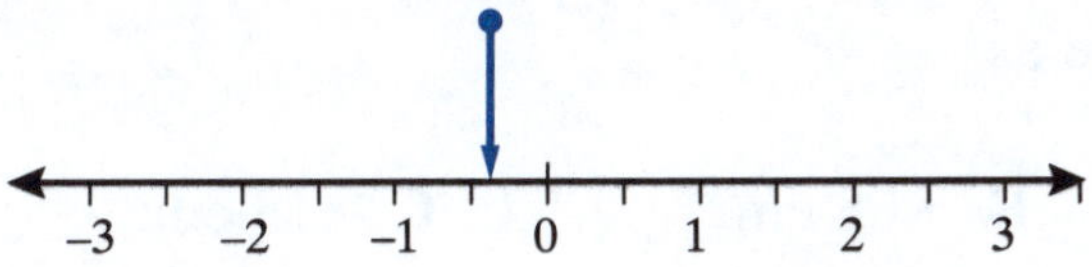


What number is at that position?

A –0.4 B –0.6 C –1.4 D –1.6

---

**Question 16**

What is the perimeter of this shape?

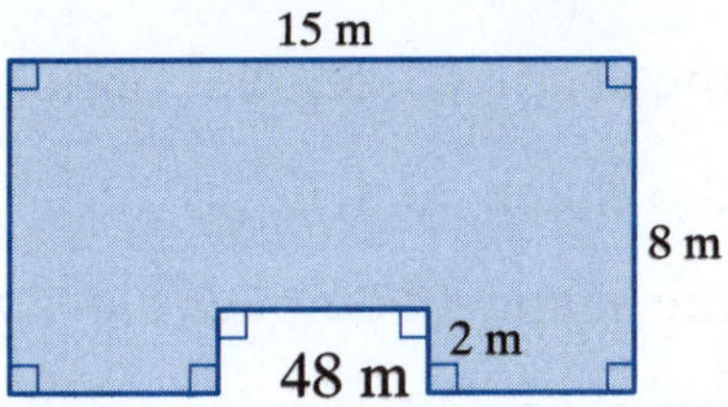


A 25 m B 46 m 48 m D 50 m

---

**It would be a good idea to check your answers to questions 9 to 16 before moving on to the other questions.**

**Question 17**

A tin holds 25 white, 12 blue, 15 black and 8 red buttons.

Without looking, Ayano takes a button from the tin.

What is the chance that it is blue?

**A** $\frac{1}{3}$ **B** $\frac{1}{4}$ **C** $\frac{1}{5}$ **D** $\frac{1}{6}$

**Question 18**

Ali made this 3D object from 12 identical cubes.

What is the view from the back?

**A** **B** **C** **D**

**Question 19**

1 cm and 50 mm is the same as

**A** 6 cm. **B** 1.5 cm. **C** 15 cm. **D** 15 mm.

**Question 20**

This shape is following a pattern.

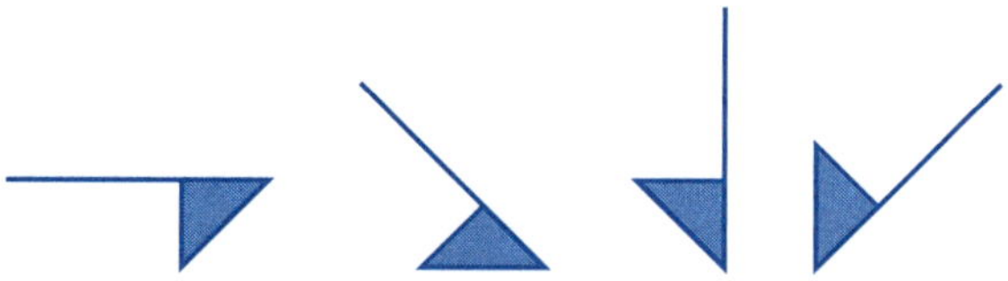

What is the rule?

**A** Rotate the shape 45 degrees anticlockwise.

**B** Rotate the shape 30 degrees anticlockwise.

**C** Rotate the shape 45 degrees clockwise.

**D** Rotate the shape 30 degrees clockwise.

**Question 21**

April travelled 32 km in half an hour. What was her average speed in kilometres per hour?

**A** 16 **B** 40 **C** 64 **D** 80

**Question 22**

One-quarter of the capacity of a bottle is 750 mL. What is the capacity of the bottle in litres?

[ ] L

**Question 23**

A car uses 7 litres of petrol for every 100 km travelled.

Which is the best estimate for the amount of petrol (in litres) needed to travel 590 km?

**A** 40 **B** 50 **C** 60 **D** 70

**Question 24**

Pretty made a 3D object from identical cubes. She drew a top view and a front view of her object.

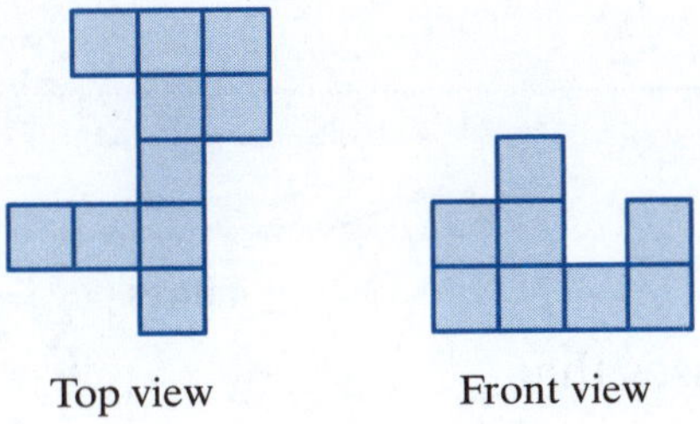


Which **cannot** be a side view?

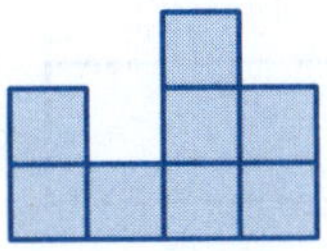

**A**

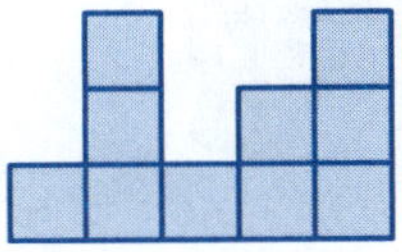

**B**

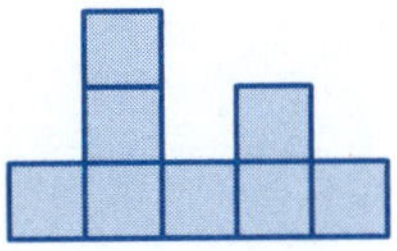

**C**

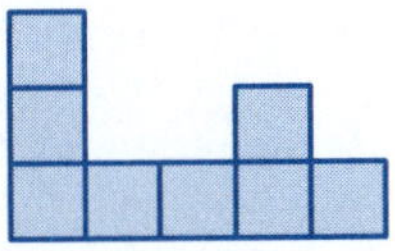

**D**

**Question 25**

Place the numbers 0.06, 0.15, 0.0087, 0.123 in order from lowest to highest.

☐ ☐ ☐ ☐

lowest highest

**Question 26**

A regular octagon is cut into 3 pieces like this:

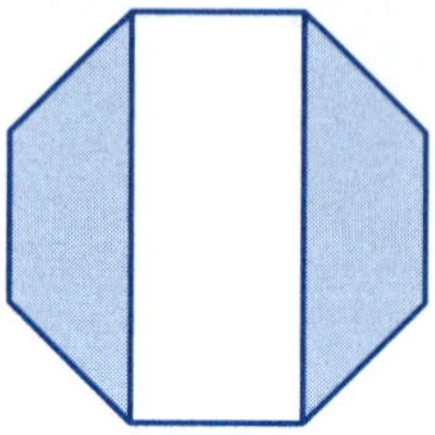

The central piece is a rectangle.

Each of the other two pieces is a

**A** rectangle. **B** pentagon. **C** hexagon. **D** rhombus. **E** trapezium.

**Question 27**

At 8:40 pm, Hui begins to watch a movie. The movie runs for 1 hour and 50 minutes.

At what time will it finish?

**A** 2030 **B** 2110 **C** 2130 **D** 2210 **E** 2230

**Question 28**

This jug has some juice in it.

From the jug, Gina fills two glasses that each hold 250 mL.

How much juice will be left in the jug?

☐ mL

**Question 29**

These numbers form a pattern but one number is missing.

1, 3, 7, 15, **?**, 63, 127, 255

What is the missing number?

☐

**Question 30**

Which spinners have a one in four chance of landing on 4? Select **all** correct answers.

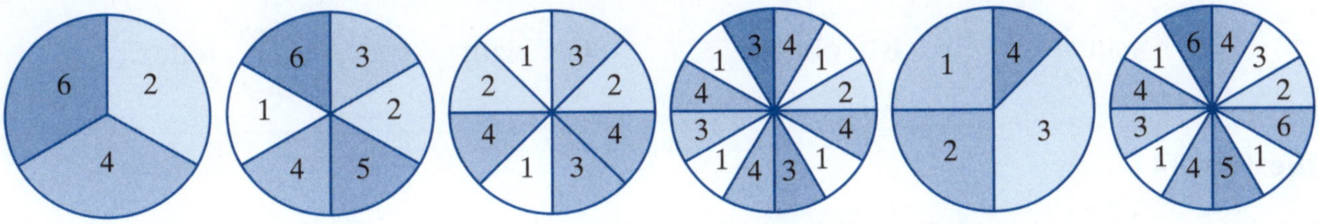


**A** **B** **C** **D** **E** **F**

**Question 31**

A butcher has 1.3 kg of mince on a tray. She sells 750 g. How many grams of the mince remain on the tray?

☐ g

**Question 32**

On a map, 1 cm represents 25 km. Two places are 3.6 cm apart on the map. What is the actual distance between the two places?

**A** 70 km  **B** 90 km  **C** 105 km  **D** 115 km

**It would be a good idea to check your answers to questions 17 to 32 before moving on.**

 Answers and explanations on pages 146–151

**Question 33**

What type of angle is the shaded angle?

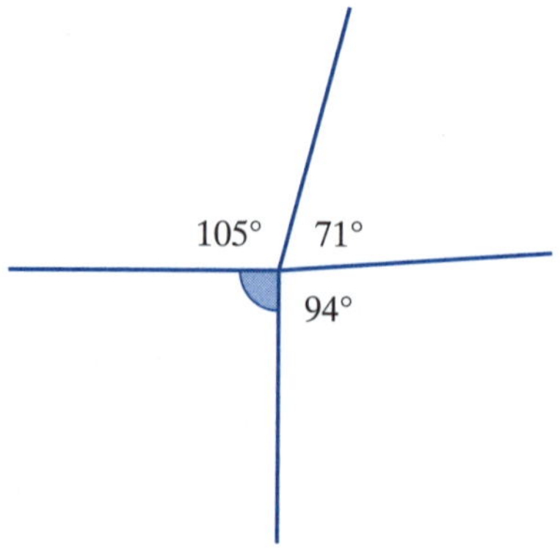


(not to scale)

**A** acute **B** obtuse **C** right **D** reflex

**Question 34**

A cube has sides of length 7 cm. What is the total area of all the surfaces?

☐ $cm^2$

**Question 35**

Julia arrived at the station and saw this clock.
Her train departed at 12:10.

How long did Julia spend at the station?

**A** 5 min **B** 25 min **C** 1 h 12 min **D** 3 h 10 min

**Question 36**

☐ and △ stand for numbers that are related by a rule.

What could be the rule? Give **all** correct answers.

| ☐ | 2 | 3 | 5 | 8 |
| --- | --- | --- | --- | --- |
| △ | 0 | 5 | 21 | 60 |

**A** △ = 5 × ☐ − 10

**B** △ = 3 × ☐ − 6

**C** △ = ☐ × ☐ − 4

**D** △ = (☐ + 2) × (☐ − 2)

**Question 37**

The graph shows the numbers of boys and girls in each year at Juniper Junior High.

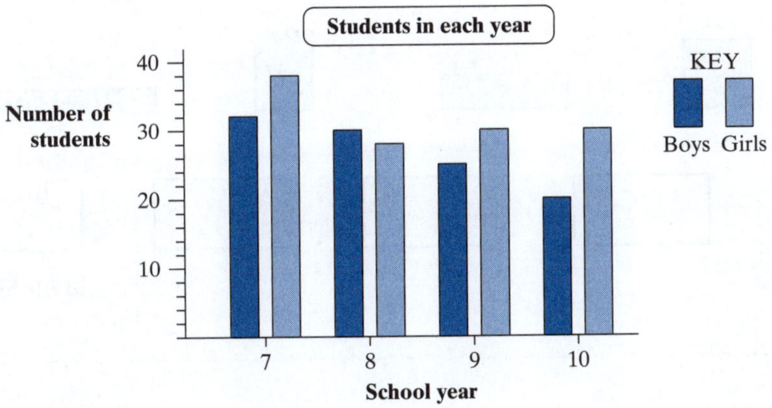


How many more girls than boys attend the school?

---

**Question 38**

There are 560 adults on a cruise. The ratio of men to women is 3 to 5. How many women are on the cruise?

**A** 210 **B** 224 **C** 336 **D** 350

---

**Question 39**

A dice has all its opposite faces adding to seven. Which could be the net of this dice? Select **all** correct answers.

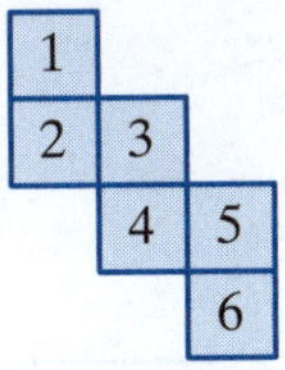


**A**

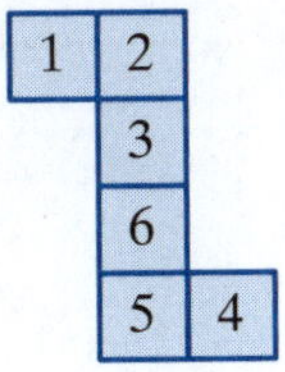


**B**

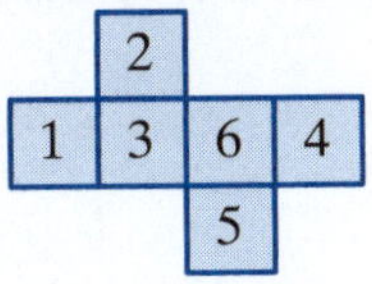


**C**

**D**

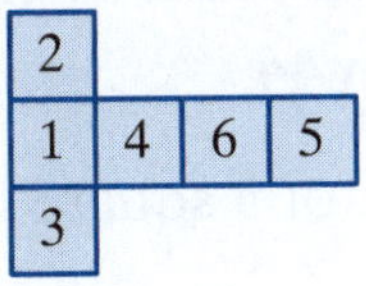


**E**

---

**Question 40**

At the fundraising event, the raffle raised $4500. This was 20% of the total amount raised. How much money was raised at the event?

**A** $900 **B** $9000 **C** $13 500 **D** $22 500


Answers and explanations on pages 146–151

**Question 41**

Write these rectangles, not drawn to scale, in order from smallest area to largest area.

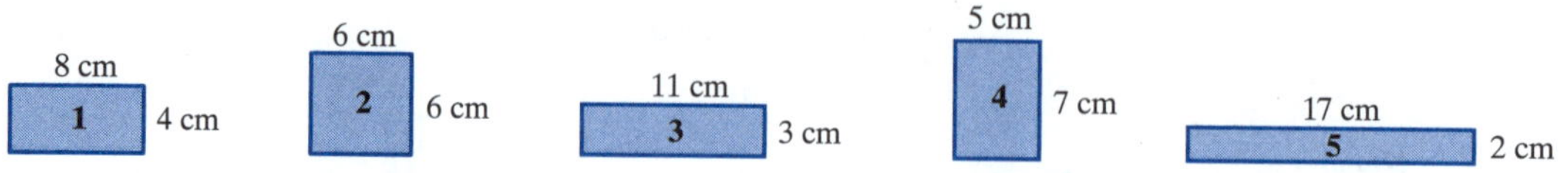


smallest area

largest area

**Question 42**

A positive number is multiplied by itself and then 7 is added. The answer is 16.

What is the number?

**Question 43**

*PS* and *RT* are straight lines. What is the size of the shaded angle?

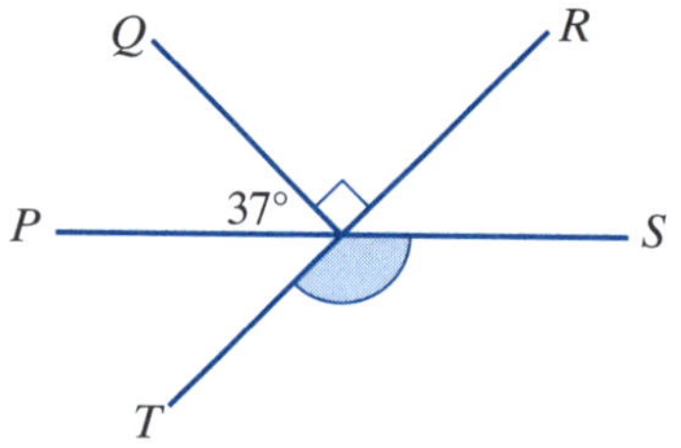


°

**Question 44**

The area of a square is 64 square metres.

What is the perimeter of the square?

m

**Question 45**

Opal used the rule 'Multiply the previous number by itself and add 5' to get the numbers in a pattern. The first number was 1. What was the fourth number?

**Question 46**

The average (mean) of Jasmine's six test scores is 65. What must she score in the seventh test to increase her average to 70?

---

**Question 47**

What number is halfway between $3\frac{3}{4}$ and $5\frac{1}{4}$?

**A** $4\frac{1}{4}$ **B** $4\frac{1}{2}$ **C** $4\frac{5}{8}$ **D** $4\frac{3}{4}$

---

**Question 48**

Write the cost of each product at the given price. Choose from these price tags:

$2.55 $2.75 $2.80 $2.90 $2.95 $3.05 $3.15 $3.25

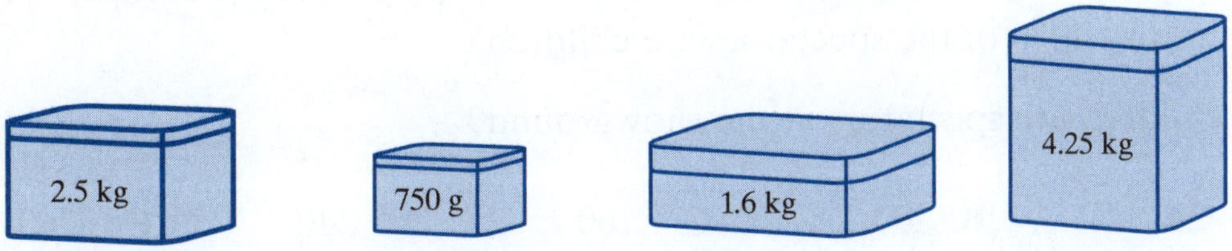


$1.30/kg $4.20/kg $1.75/kg 60c/kg

---

# END OF TEST

## Advanced level

### Section 1: Non-calculator Questions

A calculator must **not** be used for the eight questions in this part of the test.

**Question 1**

What is the next number in this pattern?

10, 9.25, 8.5, 7.75, ________

**Question 2**

There are 600 people at the showground. One-third of the people are competitors and the rest are spectators. 40% of the spectators are children.

How many children are spectators at the showground?

**A** 24 **B** 80 **C** 160 **D** 240 **E** 260

**Question 3**

Which number sentences will give the shaded area in square centimetres?
Select **all** correct answers.

**A** $37 \times 26 - 18 \times 15$
**B** $(37 - 18) \times (26 - 15)$
**C** $37 \times 26 - 11 \times 19$
**D** $37 \times 15 + 18 \times 26$
**E** $26 \times 18 + 19 \times 15$

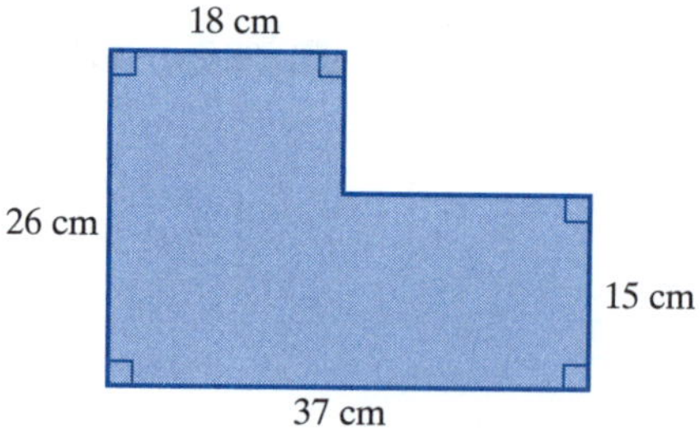


**Question 4**

What fraction is halfway between $\frac{2}{5}$ and $\frac{3}{5}$?

**Question 5**

What is the next number in this pattern? $\frac{1}{12}, \frac{1}{6}, \frac{1}{4}, \frac{1}{3}, \ldots$

**A** $\frac{1}{2}$ **B** $\frac{2}{3}$ **C** $\frac{3}{8}$ **D** $\frac{5}{12}$ **E** $\frac{1}{5}$

**Question 6**

A test has questions on perimeter, area and volume. $\frac{2}{5}$ of the test is on area and $\frac{1}{3}$ is on volume. What fraction of the test is on perimeter?

**A** $\frac{5}{8}$ **B** $\frac{4}{5}$ **C** $\frac{3}{8}$ **D** $\frac{1}{5}$ **E** $\frac{4}{15}$

**Question 7**

Write the decimal equivalent for each fraction. Choose from: 0.6, 0.75, 0.85, 0.7 or 0.8.

$\frac{3}{4}$ $\frac{4}{5}$ $\frac{2}{3}$ $\frac{7}{10}$ $\frac{17}{20}$

**Question 8**

Write the cost per litre for each amount at the given price. Choose from: \$1.65/L, \$1.70/L, \$1.75/L or \$2.00/L.

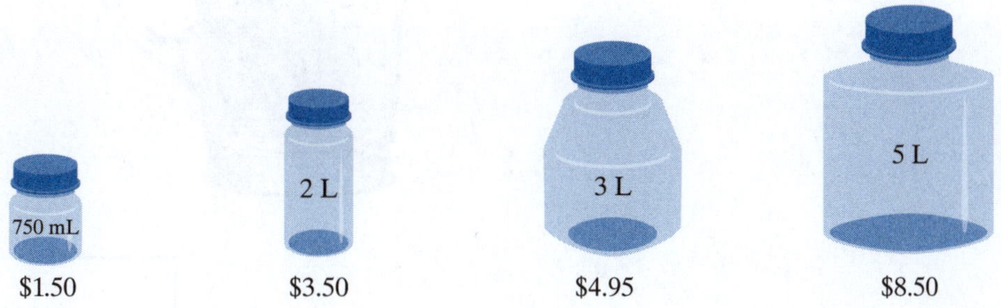


**This is the end of the part where you are not allowed to use a calculator.**
**It would be a good idea to check your answers to the questions in this section before moving on to the other questions.**

**Section 2: Calculator Allowed Questions**

**Question 9**

Isabel is standing in the garden facing West. She turns through 225° in a clockwise direction.

What feature is she now facing?

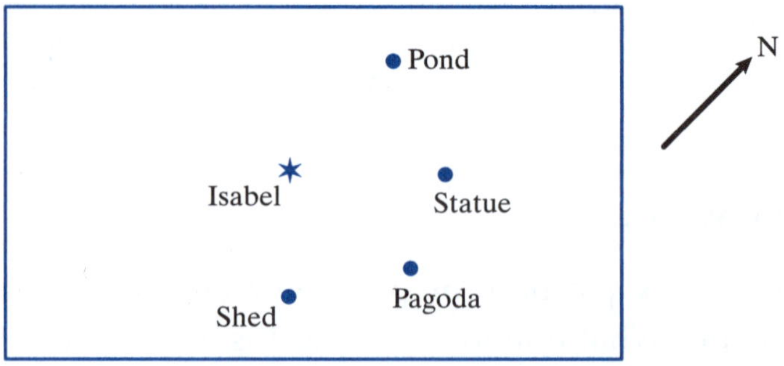


**A** Pond **B** Statue **C** Pagoda **D** Shed

**Question 10**

When 9 is added to a certain number and the result multiplied by 5, the answer is 185.

What was the number?

**Question 11**

This jug has some milk in it.
If Montana uses 600 mL of the milk, how much will remain in the jug?

mL

**Question 12**

On a map, 1 cm represents 15 km. The distance between Clifton and Underwood is actually 72 km. How far apart will they appear on the map?

cm

**Question 13**

$\frac{2}{3}$ of a number is 72. What is $\frac{3}{4}$ of the number?

**Question 14**

A committee has 2 students from Year 7, 3 from Year 8, 4 from Year 9 and 6 from Year 10. If the name of one of these students is picked from a hat, what is the chance that the student will be from Year 8?

**A** 1 in 3 **B** 1 in 4 **C** 1 in 5 **D** 1 in 8

**Question 15**

Lydia makes necklaces that she sells at a market for $8 each. The cost, in dollars, to make the necklaces is found by the rule: 'Cost = 5 × the number of necklaces + 25'.

How much profit does Lydia make if she sells 10 necklaces?

**A** $5 **B** $25 **C** $75 **D** $80

**Question 16**

When it is 8:30 am in Brisbane, it is 2:30 pm the previous day in Los Angeles. What time is it in Brisbane when it is 8:30 am Monday in Los Angeles?

**A** 2:30 am Monday **B** 2:30 am Tuesday

**C** 2:30 pm Sunday **D** 2:30 pm Monday

**It would be a good idea to check your answers to questions 9 to 16 before moving on to the other questions.**


Answers and explanations on pages 151–158

**Question 17**

Jimmy uses a travelling irrigator to water his farm. One night he paces out the distance the irrigator has moved across a paddock. He finds that it has moved 240 paces in 9 hours. Jimmy then paces out the distance the irrigator still has to move to the end of the paddock and finds it to be 160 paces. If the irrigator starts moving at 10 pm, at about what time will it reach the end of the paddock?

**A** 3 am **B** 4 am **C** 5 am **D** 6 am **E** 7 am

**Question 18**

This clock shows 4 o'clock.
What is the size of the obtuse angle formed by its hands?

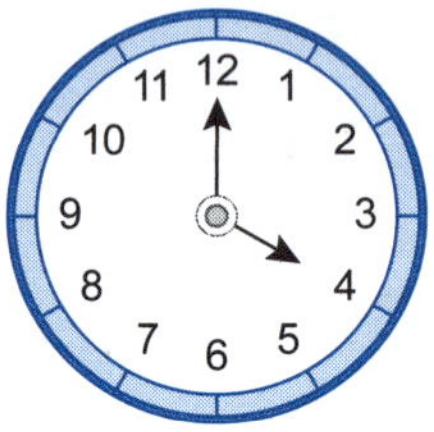


☐ °

**Question 19**

9.23 − 4.65 = ☐ + 3.18

What number will make this number sentence true? ☐

**Question 20**

This octagonal prism needs painting. The sides of the regular octagon are all 5 cm long and the prism is 9 cm wide. The area of the front face is about 120 cm$^2$.

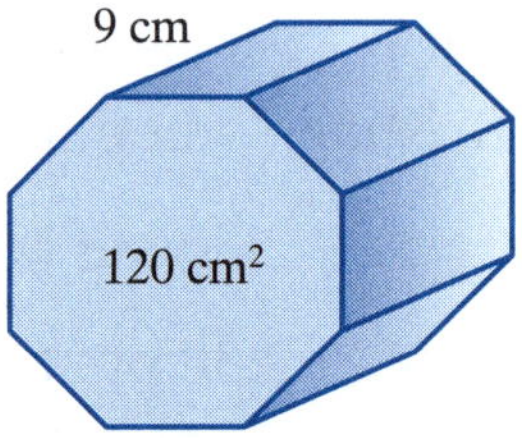


Which is the best approximation for the total area (in cm$^2$)?

**A** 400 **B** 500 **C** 600 **D** 700 **E** 800

**Question 21**

A box holds 60 ribbons. 16 of the ribbons are pink.

The fraction of pink ribbons in the box is closest to

**A** $\frac{1}{3}$ **B** $\frac{1}{4}$ **C** $\frac{1}{5}$ **D** $\frac{1}{6}$

### Question 22

Mei built an object with identical cubes. She drew three different views of it.

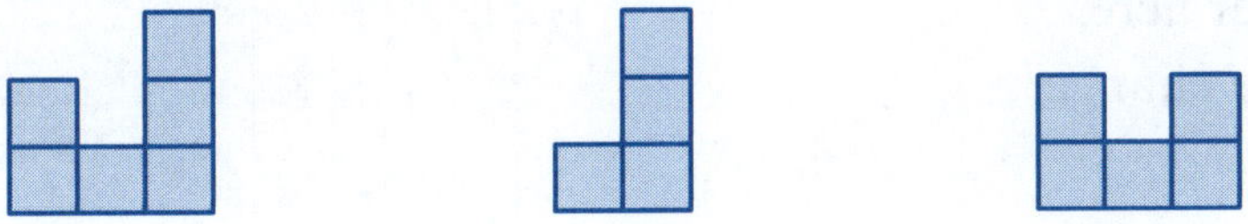

Front view Right side view Top view

Which could be the object Mei built?

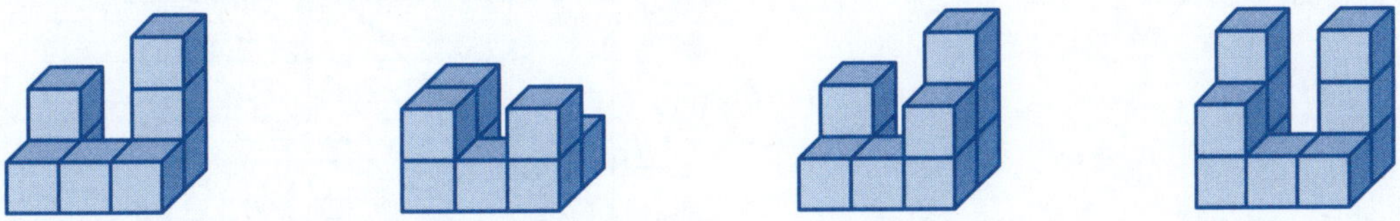

**A** **B** **C** **D**

### Question 23

A cyclist rode from Eaglevale to Sandy Beach in 45 minutes. Her average speed was 16 kilometres per hour. How far is it from Eaglevale to Sandy Beach?

[ ] km

### Question 24

This figure is made up of a square and 2 equilateral triangles.

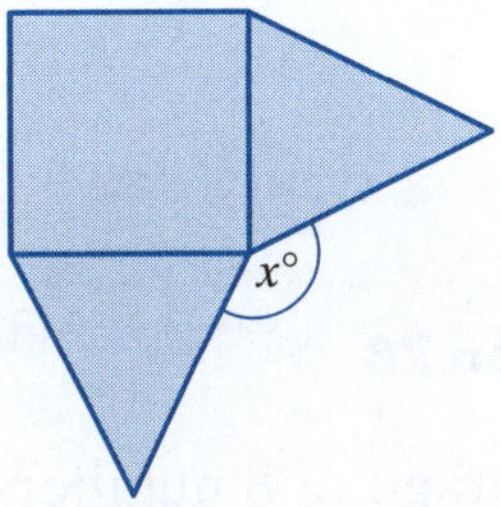


What is the value of $x$ in the diagram?

**A** 110 **B** 120 **C** 140 **D** 150 **E** 210

### Question 25

The volume of a rectangular prism is 240 cm$^3$. If the length and width are halved, but the height is doubled, what will be the volume of the new prism?

[ ] cm$^3$

**Question 26**

Wei wants to spread some fertiliser at the rate of 50 kg per acre. Wei uses this graph to change acres to hectares.

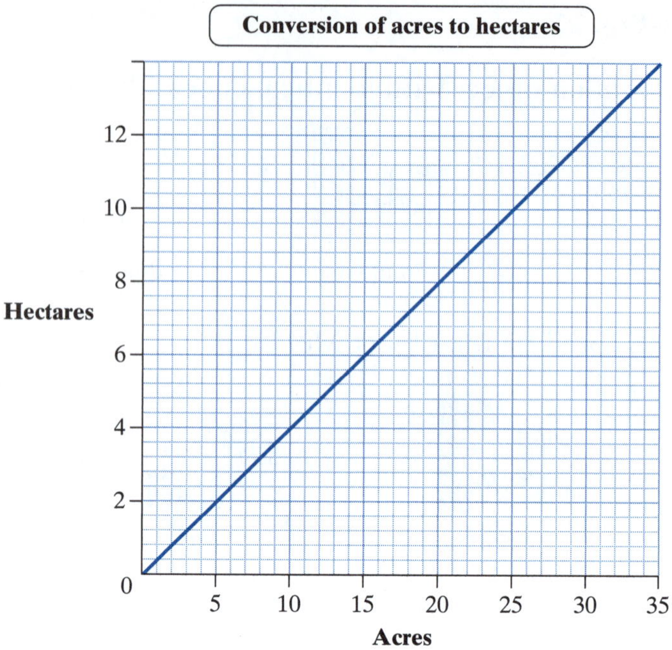


About how many kilograms of fertiliser does he need for a 3-hectare paddock?

**A** 150 **B** 375 **C** 745 **D** 1000

**Question 27**

Dane's car uses petrol at the rate of 8 litres per 100 km travelled.

How far would Dane's car travel on a full tank of 70 litres of petrol?

☐ km

**Question 28**

The average of 8 numbers is 12. If another number is included, the average increases to 14. What number is included?

**A** 14 **B** 18 **C** 28 **D** 30

**Question 29**

The diagram shows the route of a fun run that is 15 km long.

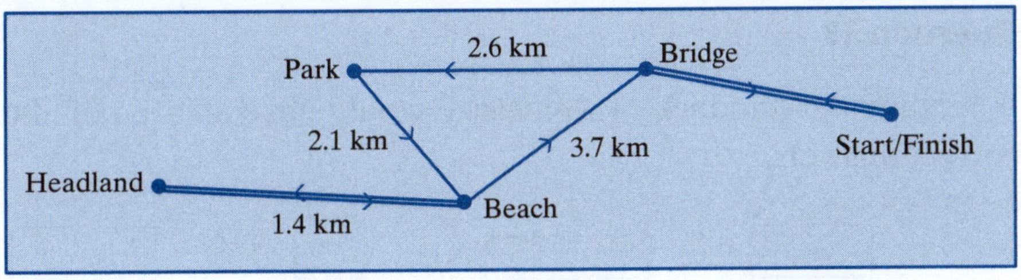


What is the distance from the bridge to the finish?

[ ] km

**Question 30**

A rule to find the correct dose of a particular medicine for a child, when the mass of the child in kilograms is known, is:

Dose for child = (Dose for adult × mass of child) ÷ 70.

What is the correct dose for a child of mass 28 kg if the adult dose is 15 mL?

[ ] mL

**Question 31**

Which of these rectangular prisms have the same volume? (All measurements are in centimetres.) Select all correct answers.

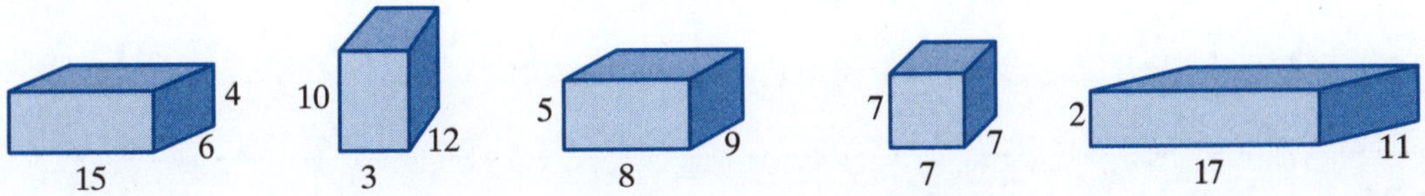


**A** **B** **C** **D** **E**

**Question 32**

Klayjian bought 4 sandwiches and a drink for $7.00. The drink cost $1.80.
What would 3 sandwiches and 2 drinks cost?

**A** $7.50 **B** $7.00 **C** $6.50 **D** $6.00

**It would be a good idea to check your answers to questions 17 to 32 before moving on.**

### Question 33

A rectangular paddock of which size would have an area of about 4 hectares? Select **all** correct answers.

(not to scale)

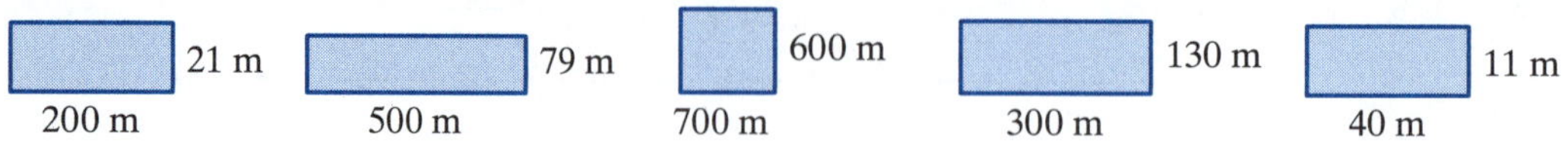


**A** **B** **C** **D** **E**

---

### Question 34

The diameter of a 10-cent coin is about 24 mm. Tanvi has $17.50 in 10-cent coins. What is the best estimate for the length of the line, if the coins are lined up in a row?

**A** 4.2 m **B** 7.3 m **C** 42 m **D** 73 m

---

### Question 35

Here are two views of the same cube. Which could be the net of the cube?

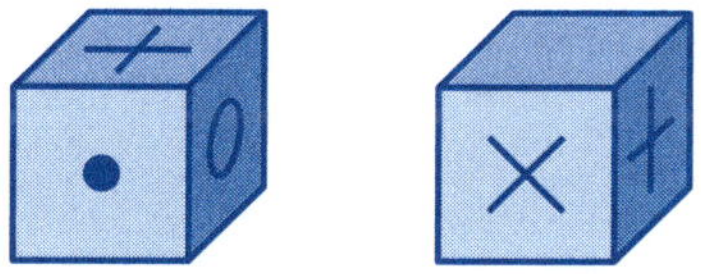

**A** **B** **C** **D**

---

### Question 36

50 people were asked which of three sporting teams they supported. Some of the results are shown in the table.

| | Crows | Eagles | Hawks | Total |
|---|---|---|---|---|
| **Men** | | 7 | 5 | 15 |
| **Women** | 8 | | 6 | |
| **Children** | 5 | 3 | | |
| **Total** | | 14 | | 50 |

How many children supported the Hawks?

**A** 4 **B** 9 **C** 11 **D** 12 **E** 13

**Question 37**

Imogen spun this spinner 40 times.

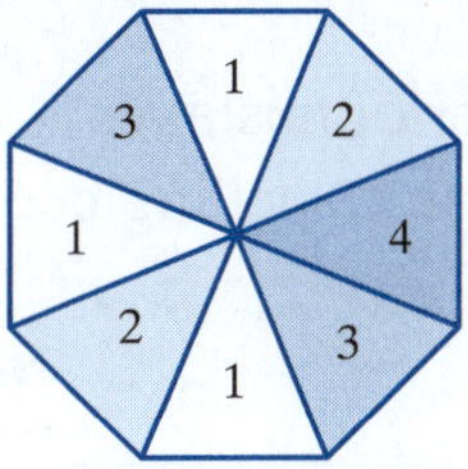


Which table is most likely to show the result?

| Score | Times |
|---|---|
| 1 | 10 |
| 2 | 10 |
| 3 | 10 |
| 4 | 10 |

**A**

| Score | Times |
|---|---|
| 1 | 5 |
| 2 | 10 |
| 3 | 10 |
| 4 | 15 |

**B**

| Score | Times |
|---|---|
| 1 | 15 |
| 2 | 10 |
| 3 | 10 |
| 4 | 5 |

**C**

| Score | Times |
|---|---|
| 1 | 13 |
| 2 | 7 |
| 3 | 7 |
| 4 | 18 |

**D**

**Question 38**

Jono's juice comes in four different sizes: small (600 mL), medium (1.5 L), large and jumbo (4.5 L). The capacity of the large size is halfway between that of the medium and the jumbo sizes. How much greater is the capacity of the large size than the small size?

☐ L

**Question 39**

The ratio of men to women at a conference was 5 to 3. There were 40 people at the conference. Later, 15 more men and 5 more women joined the conference.

What is the ratio of men to women now?

**A** 5 to 3 **B** 2 to 1 **C** 3 to 1 **D** 3 to 2 **E** 5 to 2

**Question 40**

Eason made a rectangular prism from identical cubes. The prism is 4 cubes long, 3 wide and 2 high. He painted all six faces of the prism and then broke the prism up into the individual cubes. What fraction of the cubes have just one face painted?

**A** $\frac{1}{12}$ **B** $\frac{1}{8}$ **C** $\frac{1}{6}$ **D** $\frac{1}{3}$

**Question 41**

The graph shows the number of mistakes made by all the contestants in a spelling bee.

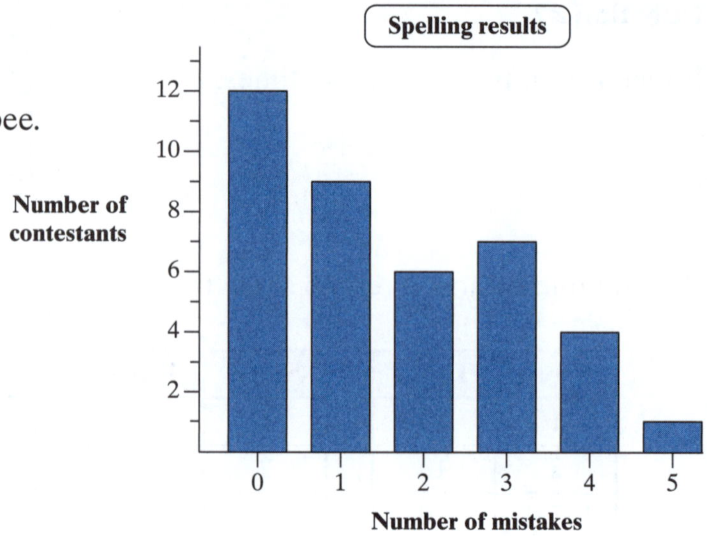


Which statements are correct?
Select **all** correct answers.

**A** 18 contestants made more than 2 mistakes.

**B** Altogether 27 mistakes were made.

**C** There were 39 contestants.

**D** No contestant made more than 5 mistakes.

**Question 42**

Place these fractions $\frac{1}{3}, \frac{3}{8}, \frac{5}{12}, \frac{7}{24}$ in order from lowest to highest.

| | | | |
|---|---|---|---|
| | | | |
| lowest | | | highest |

**Question 43**

This pattern of octagons is made with pins and string.

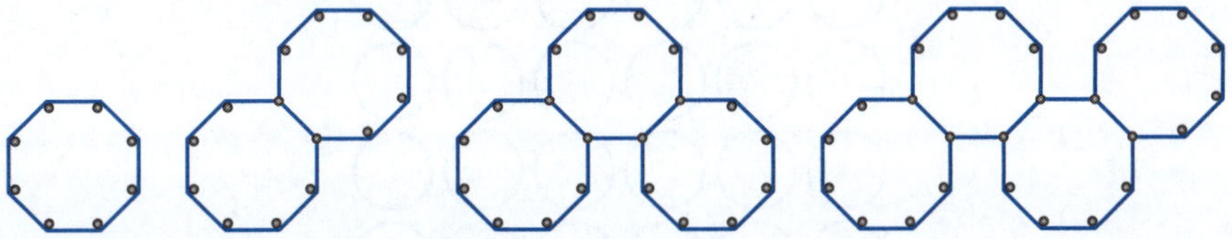

| Number of octagons | 1 | 2 | 3 | 4 |
|---|---|---|---|---|
| Number of pins | 8 | 14 | 20 | 26 |

How many octagons can be made with 200 pins?

**A** 33 **B** 34 **C** 35 **D** 36

**Question 44**

The diagram shows a regular hexagon and a square. The angles of a hexagon add to 720°.

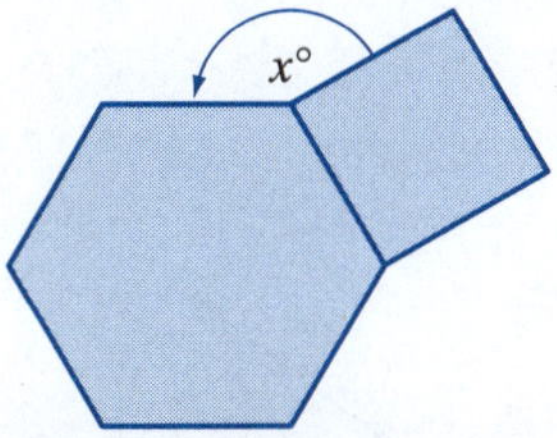


Which number sentence would give the value of $x$?

**A** $720 \div 6 + 90$ **B** $720 \div 6 - 90$ **C** $270 - 720 \div 6$ **D** $270 + 720 \div 6$

**Question 45**

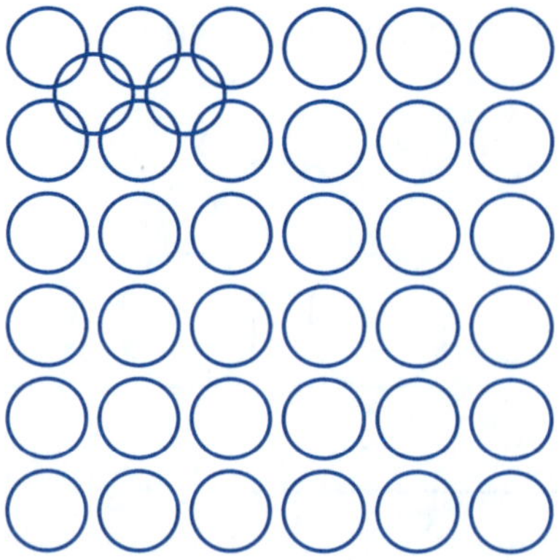

Paul is stacking cans in a display at the shop where he works. He places 6 cans across the length and 6 cans across the width at the bottom of the display, then places a second row so that each can rests on four others beneath it as shown in the diagram.

If Paul continues to stack cans in this way, adding extra rows as he goes, how many cans in total can he place in the stack?

**Question 46**

Finn made this pattern with matches.

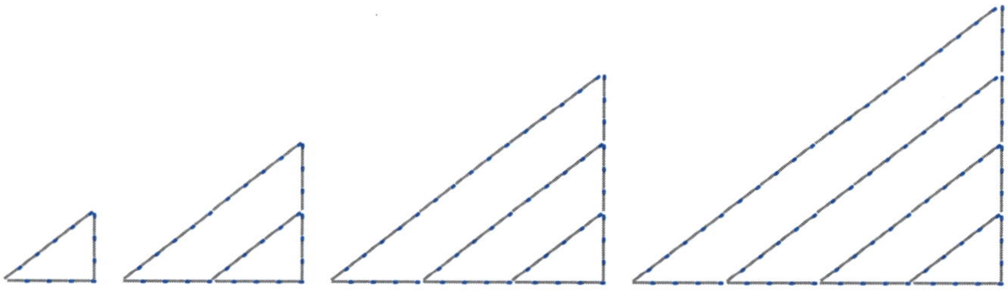

| Figure | 1 | 2 | 3 | 4 | 5 |
|---|---|---|---|---|---|
| Matches | 12 | 29 | 51 | 78 | ? |

He recorded the number of matches needed for each figure in a table.

How many matches will be needed for figure 5?

**Question 47**

This graph shows the results of an election in which Akmal got 240 votes and Dave received twice as many votes as Ali. Which statements are correct? Select **all** correct answers.

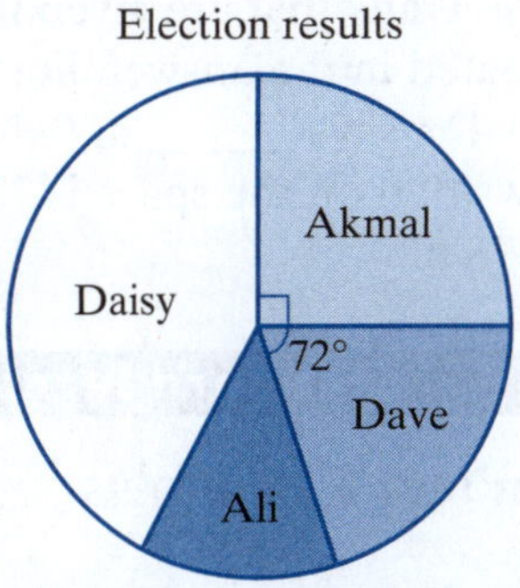


**A** There were 960 votes altogether.

**B** Akmal got exactly three times as many votes as Ali.

**C** Ali got one-tenth of all the votes.

**D** The angle size for Daisy was 162°.

**E** Dave got 202 votes.

---

**Question 48**

The town of Linden lies between Woodville and Hamlet.

Alice saw this sign on the roadside as she drove toward Linden. She knows that the distance from Woodville to Linden is twice that from Linden to Hamlet. How far is Alice from Linden?

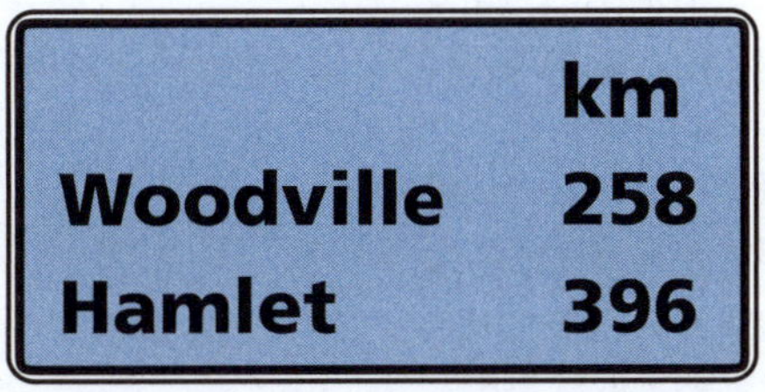


[ ] km

---

## END OF TEST

## ANSWERS TO MINI TESTS

Note: Units that are given in the question are repeated in the answer. For example, if the question says ______ m (where the number to be written on the space is 78) the answer will say 78 m.

### Standard level questions

**Mini Test 1:** Numbers ................................ Page 1

**1** C **2** 10 **3** A **4** 1000 **5** D **6** B **7** C **8** 9 **9** A **10** \$8.25 **11** D **12** 0.22 **13** 100 **14** B **15** B, C **16** B **17** \$2.25 **18** C **19** D **20** 4.6 **21** A **22** D **23** C **24** 4.7 million

**1** $(6 \times 1000) + (4 \times 10) + 7$
$= 6000 + 40 + 7$
$= 6047$
[There are no hundreds.]

**2** $3 \times \square = 6 \times 5$
Now $6 \times 5 = 30$
So $3 \times \square = 30$
The missing number must be 10 because $3 \times 10 = 30$.
[The product of two numbers will have the same value if one of the numbers is doubled and the other halved.]

**3** Five thousand and twenty-eight is 5000 plus 20 plus 8.
$5000 + 20 + 8 = 5028$
[There are 5 thousands, 0 hundreds, 2 tens and 8 units.]

**4** William has 20 packets with 50 in each packet.
Number of nails $= 20 \times 50$
$= 1000$

**5** Consider each option:
$4 \times 3 = 12$
$5 + 7 = 12$
$20 - 8 = 12$
$18 \div 3 = 6$
The option that is not equal to 12 is $18 \div 3$.

**6** $\square = 3$ and $\otimes = 5$
So $\square + \otimes = 3 + 5$
$= 8$
But $\square + \otimes = \triangle + \triangle$
So $\triangle + \triangle = 8$
This means $\triangle = 4$ because $4 + 4 = 8$.

**7** There are five divisions between each 10 marked on the number line.
Each division $= 10 \div 5$
$= 2$
So the rest of the numbers on the number line can be filled in.

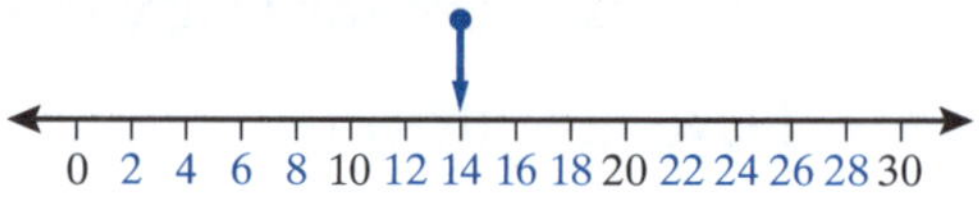


The arrow is pointing to 14.

**8** Number of remaining stickers $= 24 - 15$
$= 9$

**9** Consider each option:
$20 - 3 + 1 = 17 + 1$
$= 18$
$3 \times 5 + 1 = 15 + 1$
$= 16$
$5 + 9 + 2 = 14 + 2$
$= 16$
$6 \times 3 - 2 = 18 - 2$
$= 16$
The number sentence that is not equal to 16 is $20 - 3 + 1$.

**10** Cost of book $=$ \$7.80 + 45c
$=$ \$7.80 + \$0.45
$=$ \$8.25

$$\begin{array}{r} \$7.80 \\ + \ \$0.45 \\ \hline \$8.25 \end{array}$$

**11** $4^3$ means $4 \times 4 \times 4$.
[The index (3) tells us how many factors of 4 are multiplied together.]

**12** $0.2 + 0.02 = 0.20 + 0.02$
$= 0.22$

$$\begin{array}{r} 0.20 \\ + \ 0.02 \\ \hline 0.22 \end{array}$$

**13** $186 \div$ a number $= 1.86$
The decimal point has moved two places to the left.
1.86.0
So the number is 100.
[When dividing by powers of 10 the decimal point moves one place to the left for each

zero in the power. So when dividing by 10 the decimal point moves one place, when dividing by 100 it moves two places, and so on.]

**14** 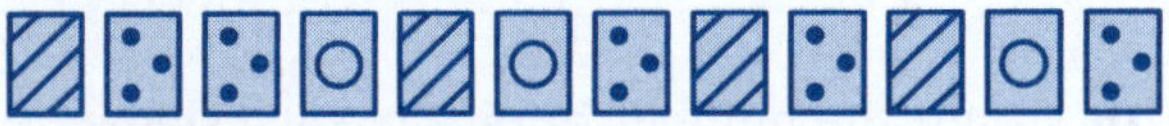

There are 12 cards altogether.
4 of the cards have stripes.
The fraction with stripes $= \frac{4}{12}$
$= \frac{1}{3}$

**15** $8 \times 4 = 32$
Now consider each option:
$10 \times 2 + 3 = 20 + 3$
$= 23$
$10 \times 3 + 2 = 30 + 2$
$= 32$
$10 \times 4 - 8 = 40 - 8$
$= 32$
$10 \times 2 \times 3 = 20 \times 3$
$= 60$
The number sentences that have the same value as $8 \times 4$ are $10 \times 3 + 2$ and $10 \times 4 - 8$.

**16** $4228 \div 7 = 604$

$$\begin{array}{r} 60\,4 \\ 7\overline{)422^{2}8} \end{array}$$

**17** Cost of ice-cream = \$5.00 − \$2.75
= \$2.25

$$\begin{array}{r} \$\,^{4}\!\not{5}.\,^{9 1}\!\not{0}\,^{1}0 \\ -\ \$\,2.\ 7\,5 \\ \hline \$\,2.\ 2\,5 \end{array}$$

**18** Population of Pinkton = 215 684
To the nearest thousand this is 216 000.
[It is closer to 216 000 than 215 000. Option A is the population rounded to the nearest hundred thousand and Option B is the population rounded to the nearest ten thousand.]

**19** Consider each option:
 has 1 of 4 equal parts shaded.
Fraction shaded $= \frac{1}{4}$
 has 1 of 4 equal parts shaded.
Fraction shaded $= \frac{1}{4}$
 has 2 of 8 equal parts shaded.
Fraction shaded $= \frac{2}{8}$
$= \frac{1}{4}$
 has 1 of 4 parts shaded but the parts are not equal in size.
Fraction shaded is not $\frac{1}{4}$.
The option that does not show $\frac{1}{4}$ shaded is Option D.

**20** $5 - 0.4 = 5.0 - 0.4$
$= 4.6$

$$\begin{array}{r} ^{4}\!\not{5}.\,^{1}0 \\ -\ 0\,.\,4 \\ \hline 4\,.\,6 \end{array}$$

**21** $17 \times 35 + 83 \times 35 = (17 + 83) \times 35$
$= 100 \times 35$
[There are 17 lots of 35 plus 83 lots of 35 so altogether there are 100 lots of 35.]

**22** 20% is one-fifth.
One-fifth of 40 = $40 \div 5$
$= 8$
8 pens are red.

**23** 79 cents is a little bit less than 80 cents.
Approximate cost = $4 \times 80$c
= 320c
= \$3.20
So the amount Walter paid will be a little bit less than \$3.20.
Of the options, the closest amount is \$3.

**24** Difference in millions = 8.2 − 3.5
= 4.7

$$\begin{array}{r} ^{7}\!\not{8}.\,^{1}2 \\ -\ 3.\ 5 \\ \hline 4.\ 7 \end{array}$$

The difference is 4.7 million.

## Standard level questions

**Mini Test 2:** Patterns ........................................ Page 2

**1** 150 **2** D **3** 25 **4** 16 **5** 19 **6** 1001 **7** A **8** 95
**9** 64 **10** 19 **11** 22 **12** 23 **13** B **14** C **15** A, C

**1** 25 50 75 100 125
The numbers are increasing by 25 each time.
[50 − 25 = 25, 75 − 50 = 25,
100 − 75 = 25, 125 − 100 = 25]
The next number is 125 + 25 = 150.

**2** 127, 119, 111, 103, 95, ….
127 − 119 = 8
119 − 111 = 8
111 − 103 = 8
103 − 95 = 8
The rule is to decrease each number by 8.

**3** Third number = 11
Fourth number = 2 × 11 + 3
= 22 + 3
= 25

**4** The second number is always twice the first number.

| First number | 1 | 2 | 3 | 4 | 8 |
|---|---|---|---|---|---|
| Second number | 2 | 4 | 6 | 8 | ? |

When the first number is 8, the second number is 2 × 8 or 16.
The missing number is 16.

**5** Count the number of matches in each figure.

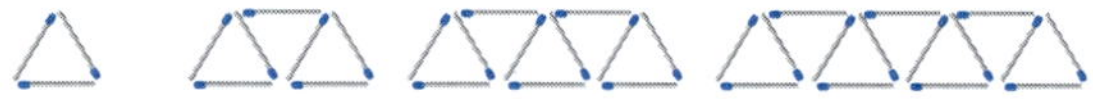

Figure 1 Figure 2 Figure 3 Figure 4

| Figure | 1 | 2 | 3 | 4 |
|---|---|---|---|---|
| Number of matches | 3 | 7 | 11 | 15 |

The number of matches increases by 4 each time.
Number of matches for Figure 5 = 15 + 4
= 19

**6** 993, 995, 997, 999, …
The numbers are increasing by 2 each time.
Next number = 999 + 2
= 1001

**7**

| | | | | | | | | | |
|---|---|---|---|---|---|---|---|---|---|
| 1 | 2 | 3 | 4 | 5 | 6 | 7 | 8 | 9 | 10 |
| 11 | 12 | 13 | 14 | 15 | 16 | 17 | 18 | 19 | 20 |
| 21 | 22 | 23 | 24 | 25 | 26 | 27 | 28 | 29 | 30 |
| 31 | 32 | 33 | 34 | 35 | 36 | 37 | 38 | 39 | 40 |

The shaded numbers are 3, 10, 17 and 24.
They are increasing by 7 each time.
Next number = 24 + 7
= 31

**8** The pattern is to subtract 10.
Next number = 105 − 10
= 95

**9** 1, 2, 4, 8, 16
The pattern is that the numbers are doubling each time.
Sixth number = 2 × 16
= 32
Seventh number = 2 × 32
= 64

**10** Count the number of pins in each shape.

Shape 1 Shape 2 Shape 3

| Shape | 1 | 2 | 3 |
|---|---|---|---|
| Number of pins | 7 | 10 | 13 |

The number of pins is increasing by 3 each time.
Number of pins for Shape 4 = 13 + 3
= 16
Number of pins for Shape 5 = 16 + 3
= 19

**11** 1, 2, 4, 7, 11, 16
The numbers increase by 1, then 2, then 3, then 4 and then 5.
The next increase will be 6.
The next number = 16 + 6
= 22

**12**

| First number | 1 | 2 | 3 | 4 | 5 |
|---|---|---|---|---|---|
| Second number | 7 | 11 | 15 | 19 | ? |

The first numbers increase by 1 each time.
The second numbers increase by 4 each time.
When the first number is 5, the second number will be 19 + 4 or 23.
The missing number is 23.

**13** 7, 18, 29, 40, …
18 − 7 = 11
29 − 18 = 11
40 − 29 = 11
The numbers are increasing by 11 each time.

**14** The first number is 5.
The second number is 5 + 6 [or 5 plus 1 lot of 6.]
The third number is 5 + 2 × 6.
The fourth number is 5 + 3 × 6.
[So each number is 5 plus a certain multiple [of 6. This multiple is 1 less than the position of the number.]
Tenth number = 5 + 9 × 6
= 5 + 54
= 59
[Or just write out all the numbers until there are ten: 5, 11, 17, 23, 29, 35, 41, 47, 53, 59.]

**15** The only information you have from the table is that the second number is 7 when the first number is 4.

| First number | 1 | 2 | 3 | 4 | 5 |
|---|---|---|---|---|---|
| Second number | | | | 7 | |

Try each option:
'Add 3 to the first number.'
$4 + 3 = 7$ ✓
This could be the rule.
'Subtract the first number from 10.'
$10 - 4 = 6$ ✗
This could not be the rule.
'Double the first number and subtract 1.'
$2 \times 4 - 1 = 8 - 1$
$= 7$ ✓
This could be the rule.
'Multiply the first number by 3 and add 4.'
$3 \times 4 + 4 = 12 + 4$
$= 16$ ✗
This is not the rule.
The rule could be 'Add 3 to the first number' or 'Double the first number and subtract 1'.

## Standard level questions

**Mini Test 3:** Geometry................................ Page 3

**1** C **2** D **3** 1 **4** B **5** A **6** 7 **7** 15 **8** 10 **9** C
**10** A **11** C **12** A **13** D **14** B **15** B **16** D

**1** The object has a constant circular cross-section.
It is a cylinder.

**2** The shape has 4 sides so it is a quadrilateral (not a pentagon which has 5 sides). It has one pair of opposite sides parallel so it is a trapezium.

[A parallelogram has both pairs of opposite sides parallel and a rhombus is a parallelogram with all of its sides equal.]

**3** The faces 2, 4, 5 and 6 will all be next to face 3. The face that is not next to 3 is the one that is opposite it. It is the face numbered 1.

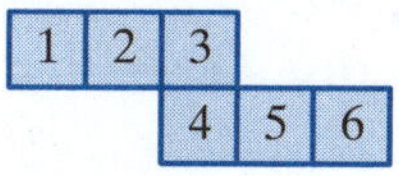


**4** The net has 4 triangular faces. It is the net of a triangular pyramid (sometimes known as a tetrahedron).

**5** The pentagon has no parallel sides.

[B is a trapezium; it has one pair of parallel sides. C is a regular hexagon; it has three pairs of parallel sides. D is a regular octagon; it has four pairs of parallel sides.]
The option with no parallel sides is A.

**6** A rectangular prism has 6 faces. This shape has 1 extra face. It has 7 faces.

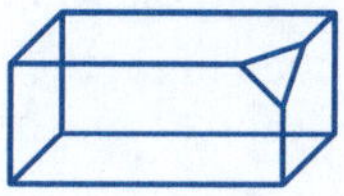

**7** A rectangular prism has 12 edges. This shape has 3 extra edges. It has 15 edges.

**8** A rectangular prism has 8 vertices. This shape has 2 extra vertices.
[1 vertex was removed and was replaced by 3 new ones.]
It has 10 vertices.
[Note that Euler's result (the sum of the faces and vertices is 2 more than the number of edges) still holds.]

**9** Consider each option:

A is bigger than a right angle. It measures more than 90°.

B is a right angle. It measures 90°.

C measures about 45°.

D measures about 20°.

The angle that measures about 45° is C.

**10** Consider each option:

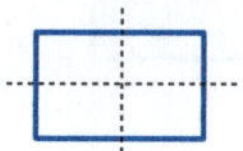

A rectangle has 2 lines of symmetry.

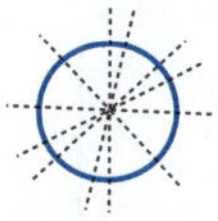

A circle has an infinite number of lines of symmetry.

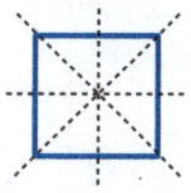

A square has 4 lines of symmetry.

# Year 7 Numeracy Mini Test Answers

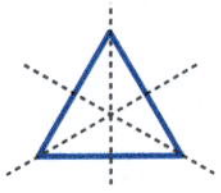

An equilateral triangle has 3 lines of symmetry.

The shape that has exactly two lines of symmetry is the rectangle.

**11** [The shape has 3 cubes in the front row and 7 cubes in the back row. When viewed from the top there will be 2 rows of squares. The back row has squares 4 across and the front row has only 2 squares across.]

The view from the top is 

**12** [From the right side there will only be 2 rows of squares. Because the back row is 3 cubes high (at most) and the front row 2 cubes high, the view from the right will show 2 squares on the left and 3 squares on the right.]

The view from the right side is 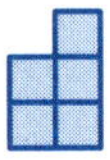

**13** [From the back only the cubes in the back row will be seen.]

The view from the back is 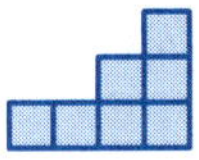

**14** From the post office the school is south.

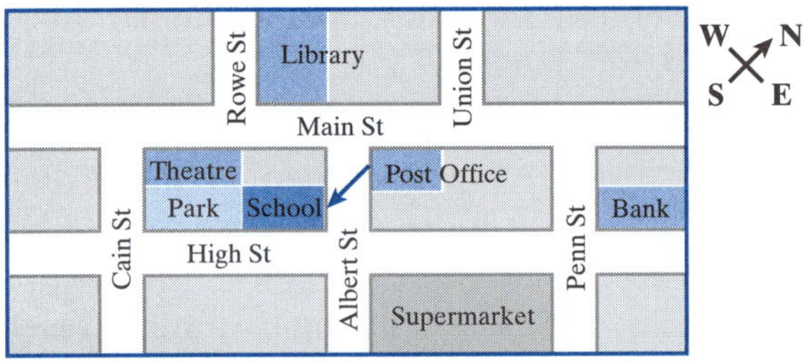


**15** Gavin is in Rowe Street.

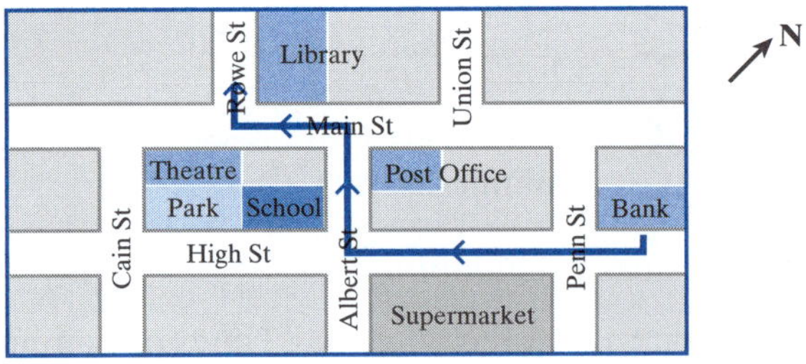


**16** Serafina is facing Cain Street.

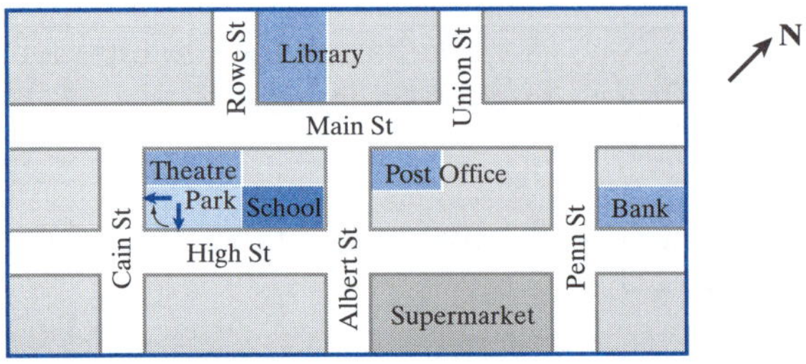


## Standard level questions

**Mini Test 4:** Numbers and Patterns ...... Page 4

**1** 12 kg **2** D **3** 32 **4** 1900 **5** 130 **6** B **7** \$2.40 **8** 10 **9** 33 **10** \$43 **11** 7880 L **12** \$37.30 **13** B **14** 36 **15** A **16** 7 513 000 **17** B, D **18** C **19** 64 **20** \$15 **21** 5.4 **22** 171 **23** 272

**1** [Each kilogram cost \$2.50 so dividing the total cost by \$2.50 will give the number of kilograms.]
Number of kilograms = \$30 ÷ \$2.50
= 12

**2** Number of years = 2020 − 1885
= 135

**3** [Sue received \$24 for every vest she knitted, so dividing the total amount by \$24 will give the number of vests.]
Number of vests = \$768 ÷ \$24
= 32

**4** [The average is found by dividing the total number of marks by the number of students. So the total marks will be equal to the average multiplied by the number of students.]
Total marks = 76 × 25
= 1900

**5** 5% of the houses have a pool.
Number of pools = 5% of 2600
$= \frac{5}{100} \times 2600$
= 130
[10% of 2600 is 260. 5% is half of 10%. So 5% of 2600 is 260 ÷ 2 or 130.]

**6** Fraction $= \frac{18}{24}$
$= \frac{3}{4}$ (after dividing both numerator and denominator by 6)

**7** Cost of 2 sausage rolls = 2 × \$1.80
= \$3.60
Cost of 3 pies = \$10.80 − \$3.60
= \$7.20
Cost of 1 pie = \$7.20 ÷ 3
= \$2.40

**8** 38 + 53 = 81 + □
Now 38 + 53 = 91
So 81 + □ = 91
This means that □ is 10 because 81 + 10 = 91.
The required number is 10.

**9** 6534 divided by a number is 198.
Number = 6534 ÷ 198
= 33

**10** $\frac{786 - 528}{32 - 26} = \frac{258}{6}$
$= 43$
Cost per hour = $43

**11** [8250 L is 370 L more than the previous amount.]
Previous amount = 8250 L − 370 L
= 7880 L

**12** Total amount paid = $100 − $12.75
= $87.25
Cost of sheets = $49.95
Cost of the pillow = $87.25 − $49.95
= $37.30

**13** $3\frac{2}{5} = \frac{3 \times 5 + 2}{5}$
$= \frac{17}{5}$
[Every whole number can be divided into 5 fifths so the number of fifths in 3 is 3 × 5.]

**14** 88, 75, 62, 49, .....
[88 − 75 = 13, 75 − 62 = 13, 62 − 49 = 13]
The numbers are decreasing by 13 each time.
Next number = 49 − 13
= 36

**15** △ = 2 and □ = 5
△ + ○ = □ − △
Now □ − △ = 5 − 2
= 3
So 2 + ○ = 3
This means ○ = 1 because 2 + 1 = 3.

**16** 7 512 862
To the nearest 1000, the population is 7 513 000.
[It is closer to 7 513 000 than 7 512 000.]

**17** 9 × 7 = 63
Consider each option:
12 + 9 × 3 = 12 + 27
= 39
11 × 6 − 3 = 66 − 3
= 63
8 × 8 − 3 = 64 − 3
= 61
12 × 5 + 3 = 60 + 3
= 63
The expressions with the same value as 9 × 7 are 11 × 6 − 3 and 12 × 5 + 3.
[Any multiplications or divisions are done before any additions or subtractions.]

**18** $4^2 = 4 \times 4$
= 16
$6^2 = 6 \times 6$
= 36
Of the options, the number between 16 and 36 is 31.

**19** [Each person paid $8.50 so dividing the total takings by $8.50 will give the number of people.]
Number of people = $544 ÷ $8.50
= 64

**20** Each metre of material cost $6.
Harry bought 2.5 metres.
Total cost = 2.5 × $6
= $15

**21** 3, 3.6, 4.2, 4.8, …
3.5 – 3 = 0.6
4.2 – 3.6 = 0.6
4.8 – 4.2 = 0.6
The numbers are increasing by 0.6 each time.
Next number = 4.8 + 0.6
= 5.4

**22** $\frac{1}{3}$ of the animals = 57
Total number of animals = 57 × 3
= 171
There are 171 animals at the shelter.

**23** $\frac{2}{3}$ of the students are male so $\frac{1}{3}$ are female.
Number of female students = $\frac{1}{3}$ of 816
= 816 ÷ 3
= 272
There were 272 female students taking the course.

## Standard level questions

**Mini Test 5:** Measurement Page 5

**1** B **2** C **3** B **4** 38 mm **5** A **6** 9 cm **7** 36 min
**8** D **9** B **10** D **11** A **12** C **13** 140 m **14** A
**15** 5000 cm **16** 35 min **17** 180 m

**1** Speed = 300 km in 5 hours
= 60 km in 1 hour
= 60 km/h

**2** [The scale shows heights of 3 m and 2 m. The water level is below 2 m. There is one mark between 2 m and 3 m on the scale so that would represent a height of 2.5 m. This means that the other mark that can be seen below 2 m would represent 1.5 m. The water level is below 1.5 m.]

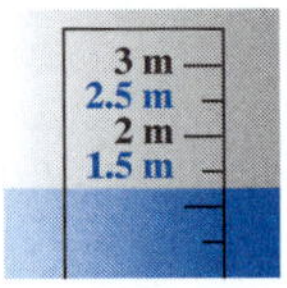


The level of water is about 1.25 m.

**3** There are 7 full squares.
The remaining parts of squares can be combined to make about another 3 squares.

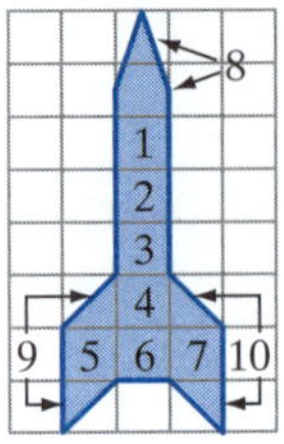


The total area is about 10 squares.

**4** The head of the nail is at 1 cm or 10 mm.
The point of the nail is at 48 mm.
Length of the nail = 48 mm − 10 mm
= 38 mm

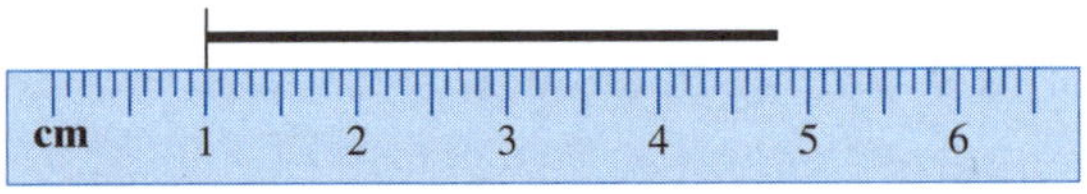


**5** [The area of any triangle is found by multiplying half the base by the perpendicular height. (Perpendicular means at right angles. The height must be at right angles to the base.) In this case, the base is 9 and the perpendicular height is 8.]
The area is found by the calculation
$\frac{1}{2} \times 9 \times 8$.

**6** [If the perimeter of a square is 36 cm, all 4 sides add to 36 cm. All the sides of a square are the same length.]
Length of each side = 36 cm ÷ 4
= 9 cm

**7** A quarter past two is 2:15 or 15 minutes after two o'clock. 2:51 is 51 minutes after two o'clock.
Difference in minutes = 51 − 15
= 36
Samuel must wait 36 minutes for the train.

**8** 7 kg = 7000 g
so 7.25 kg = 7250 g

**9** [Fill in more measurements on each of the jugs.]

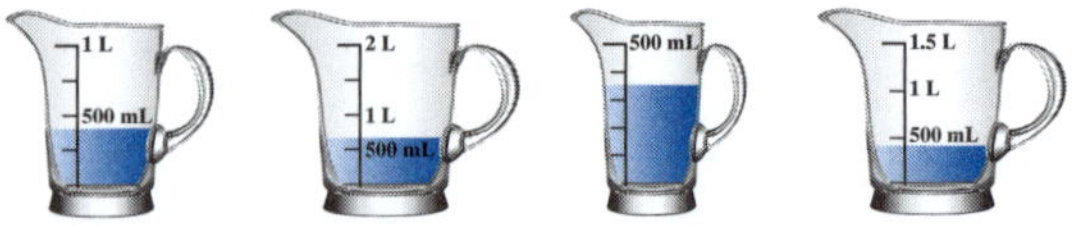


The jugs in Options A, C and D hold a little less than 500 mL.
The jug in Option B holds a little more than 500 mL.
The jug that holds the most is B.

**10** Area of rectangle = length × width
= 100 m × 50 m
= 5000 m$^2$

**11** 18:15 is in 24-hour time.
Now 18 − 12 = 6
[So 1800 h is 6 h after 12 noon.]
So 18:15 is 6:15 pm.

**12** Speed = 8 km in $\frac{1}{2}$ hour
= 16 km in 1 hour
= 16 km/h

**13** [The rectangle has two sides 50 m long and two sides 20 m long.]

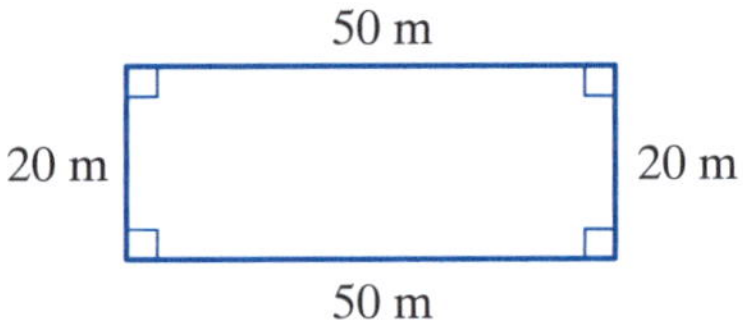


Perimeter = (50 + 20 + 50 + 20) m
= 140 m

**14** Petrol consumption = 9 L every 100 km
Now 500 km = 5 × 100 km
So petrol used = 5 × 9 L
= 45 L

**15** In 1 m there are 100 cm.
In 50 m there are 50 × 100 cm or 5000 cm.

**16** 8:40 is 20 minutes before 9 o'clock.
9:15 is 15 minutes after 9 o'clock.
Total time = (20 + 15) min
= 35 min

**17** One-third of the length is 60 m.
Total length = 3 × 60 m
= 180 m

# Year 7 Numeracy Mini Test Answers

## Standard level questions

**Mini Test 6:** Tables and Graphs................ Page 6

**1** 19 **2** 37 **3** 38 **4** 140 **5** 5 **6** A **7** A **8** A, B **9** 1750 **10** 3575 **11** 2925 **12** 29 525 **13** 16 **14** 3 **15** 82 **16** D

**1** From the table the number of girls who had travelled overseas = 19.

**2** There were 56 girls in total. 19 had travelled overseas.
Number of girls who had not travelled overseas = 56 − 19
= 37

**3** 61 boys had not travelled overseas. 23 boys had travelled overseas.
Difference = 61 − 23
= 38

**4** Total number of boys = 23 + 61
= 84
Total surveyed = 84 + 56
= 140

| | Had | Had not | Total |
|---|---|---|---|
| **Boys** | 23 | 61 | **84** |
| **Girls** | 19 | **37** | 56 |
| **Total** | **42** | **98** | **140** |

**5**

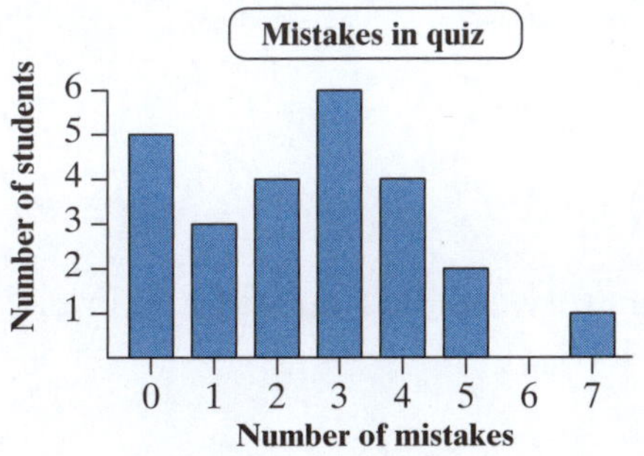


5 students made 0 mistakes.

**6** 6 students made 3 mistakes. This is the most number of people.
The most common number of mistakes is 3.

**7** 4 students made 4 mistakes.
4 students made 2 mistakes.
Difference = 4 − 4
= 0

**8** Consider each option:

1 student made 7 mistakes. This was the greatest number of mistakes. The statement 'The greatest number of mistakes was 7' is correct.

The number of students who made 6 mistakes is 0. The statement 'No student made exactly 6 mistakes' is correct.

5 students made no mistakes. 3 students made 1 mistake. Total number of students who made fewer than 2 mistakes is 5 + 3 = 8.

The statement '12 students made fewer than 2 mistakes' is not correct.

4 people made 2 mistakes. The statement '5 people made 2 mistakes' is not correct.

The statements that are correct are 'The greatest number of mistakes was 7' and 'No student made exactly 6 mistakes'.

**9** The population of Bunyip is 6550.
The population of Earlwood is 4800.
Difference = 6550 − 4800
= 1750

| Town | Population |
|---|---|
| Amity | 8700 |
| Bunyip | 6550 |
| Carrington | 4350 |
| Douglass | 5125 |
| Earlwood | 4800 |

The population of Bunyip is greater than that of Earlwood by 1750.

**10** The population of Douglass is 5125.
The population of Amity is 8700.
Difference = 8700 − 5125
= 3575
The population of Douglass is less than that of Amity by 3575.

**11** Combined population of Carrington and Douglass = 4350 + 5125
= 9475
The population of Bunyip is 6550.
Difference = 9475 − 6550
= 2925
The combined population of Carrington and Douglass is 2925 more than the population of Bunyip.

# Year 7 Numeracy Mini Test Answers

**12** Total population
$= 8700 + 6550 + 4350 + 5125 + 4800$
$= 29\,525$

**13**

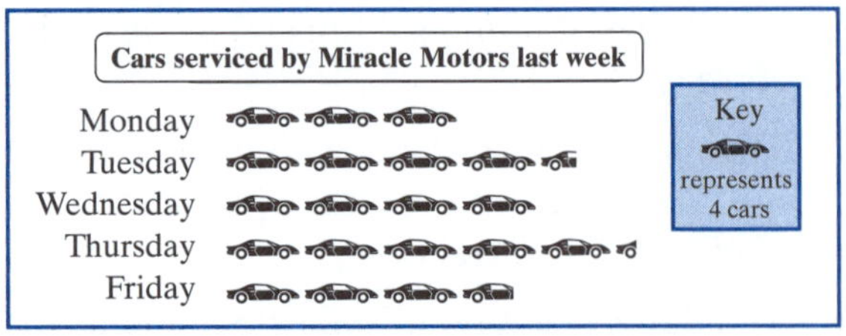


[1 symbol represents 4 cars, so half a symbol represents $\frac{1}{2}$ of 4 cars or 2 cars. One-quarter of a symbol represents 1 car and three-quarters represents 3 cars.]

On Wednesday there are 4 symbols.
Each symbol represents 4 cars.
Number of cars serviced on Wednesday
$= 4 \times 4$
$= 16$

**14** On Friday there are $3\frac{3}{4}$ symbols. On Monday there are 3 symbols.

Difference $= \frac{3}{4}$ symbol

Each symbol represents 4 cars.

$\frac{3}{4}$ of a symbol represents $\frac{3}{4}$ of 4 cars or 3 cars.

3 more cars were serviced on Friday than on Monday.

**15** Total number of symbols

$= 3 + 4\frac{1}{2} + 4 + 5\frac{1}{4} + 3\frac{3}{4}$

$= 20\frac{1}{2}$

Total number of cars $= 20\frac{1}{2} \times 4$
$= 20 \times 4 + \frac{1}{2} \times 4$
$= 80 + 2$
$= 82$

[Or count the number of full symbols: 19. This represents $19 \times 4$ cars or 76 cars. Then add the cars represented by fractional symbols: 2 on Tuesday, 1 on Thursday and 3 on Friday. The total is 82.]

**16** Consider each option:
On Thursday there are $5\frac{1}{4}$ symbols.

Number of cars $= 5\frac{1}{4} \times 4$
$= 5 \times 4 + 1$
$= 21$

The statement '21 cars were serviced on Thursday' is correct.

Number of symbols for Monday and Thursday

$= 3 + 5\frac{1}{4}$

$= 8\frac{1}{4}$

Number of symbols for Tuesday and Friday

$= 4\frac{1}{2} + 3\frac{3}{4}$

$= 8\frac{1}{4}$

The number of symbols is the same so the number of cars will be the same.
The statement 'The same number of cars were serviced on Monday and Thursday as the number on Tuesday and Friday' is correct.

Number of symbols for Wednesday $= 4$

Number of symbols for Friday $= 3\frac{3}{4}$

Difference $= \frac{1}{4}$

$\frac{1}{4}$ of a symbol represents 1 car.

The statement 'One more car was serviced on Wednesday than on Friday' is correct.

21 cars were serviced on Thursday.
Number of cars serviced on Friday

$= 3\frac{3}{4} \times 4$
$= 3 \times 4 + 3$
$= 12 + 3$
$= 15$

Total cars serviced on Thursday and Friday
$= 21 + 15$
$= 36$

The statement 'A total of 32 cars were serviced on Thursday and Friday' is incorrect.
The incorrect statement is 'A total of 32 cars were serviced on Thursday and Friday.'

## Standard level questions



**1** 506 020 **2** D **3** A **4** A **5** 9 **6** C
**7** 0.019, 0.07, 0.25, 0.4 **8** C **9** D
**10** 38 m **11** 47 **12** 2 **13** 7 **14** D **15** 6 **16** C

# Year 7 Numeracy Mini Test Answers

**1** Five hundred and six thousand and twenty
= 500000 + 6000 + 20
= 506020
[5 × 100000 + 0 × 10000 + 6 × 1000
+ 0 × 100 + 2 × 10 + 0 × 1]

**2** The shape was 4 squares across and, in total, 6 squares down.

The new shape must be
4 × 2 = 8 squares across and
6 × 2 = 12 squares down.

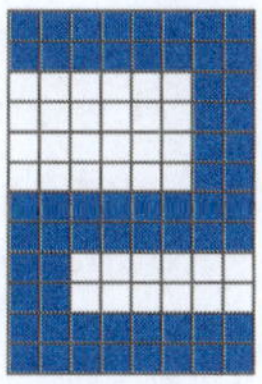

The new shape is D.

**3** The shape turns through a quarter of a turn, or 90°, in a clockwise direction.

The correct option is A.

**4** Speed = 200 km in 4 hours
= 50 km in 1 hour
Guy's average speed was 50 km/h.

**5**

| Sleeve type | Small | Medium | Large |
|---|---|---|---|
| Sleeveless | 23 | 15 | 8 |
| Short | 34 | 29 | 28 |
| Long | 21 | 24 | 14 |

There were 23 small sleeveless shirts sold and 14 long-sleeved large shirts.
Difference = 23 − 14
= 9
9 more small sleeveless shirts were sold than large shirts with long sleeves.

**6** (5 × 1000) + (2 × 100) + 6
= 5000 + 200 + 6
= 5206
[There are no tens.]

**7** Write each decimal with the same number of places after the decimal point:
0.400, 0.070, 0.019, 0.250.
So, in order, the numbers are:
0.019, 0.070, 0.250 and 0.400.
From lowest to highest the decimals are
0.019, 0.07, 0.25 and 0.4.

**8** Consider the options:
A triangular pyramid has 4 faces and 4 vertices.
A pentagonal pyramid has 6 faces and 6 vertices.
A hexagonal prism has 8 faces and 12 vertices.
An octagonal pyramid has 9 faces and 9 vertices.
Of the choices, the object that does not have the same number of faces as vertices is the hexagonal prism.

**9** [If a shape is folded along a line of symmetry the two halves must match exactly.]
Only in the circle is the dotted line a line of symmetry.

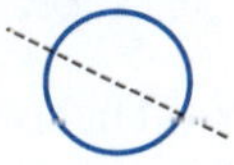

**10**

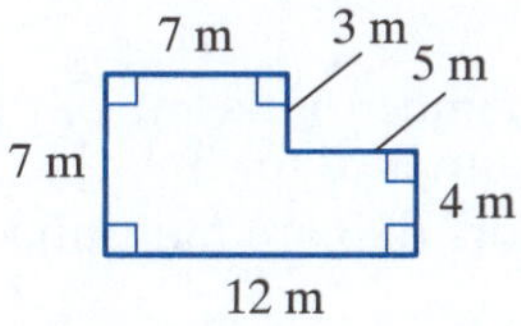


The bottom is 12 m long so the two top sections must total 12 m.
The left side is 7 m so the two right side sections must total 7 m.
Perimeter = 12 m + 7 m + 12 m + 7 m
= 38 m

**11** 15, 23, 31, 39, …
The numbers are increasing by 8 each time.
Next number = 39 + 8
= 47

**12** 32 ÷ 8 = 8 ÷ ⊗
Now 32 ÷ 8 = 4
So 8 ÷ ⊗ = 4
Now 8 ÷ 2 = 4 so ⊗ = 2.

**13** 11 women had had 3 children.
4 women had had 5 children.

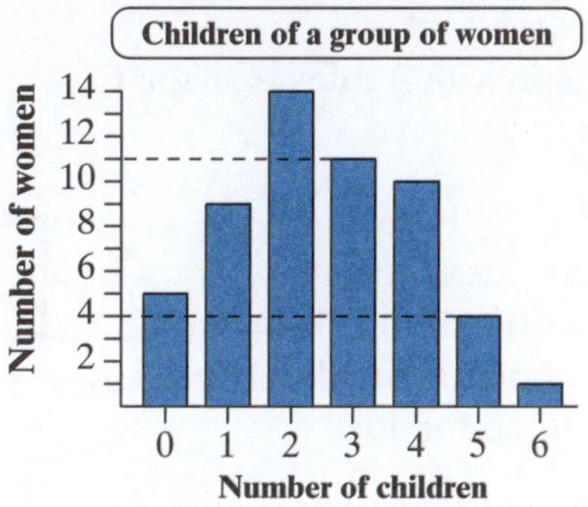


Difference = 11 − 4
= 7
There were 7 more women who had had 3 children than the number that had had 5 children.

# Year 7 Numeracy Mini Test Answers

**14** The net has 5 faces, 1 rectangular face and 4 triangular faces. It is the net of a rectangular pyramid.

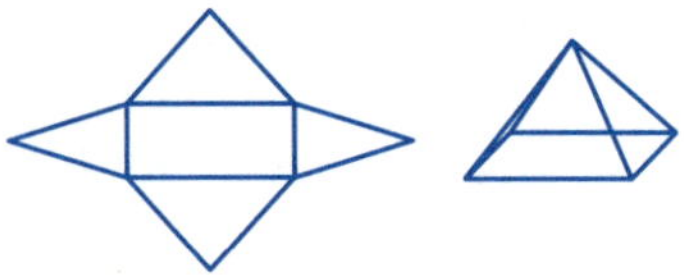

**15** 3, 5, 10
Sum = 3 + 5 + 10
= 18
There are 3 numbers.
Average = 18 ÷ 3
= 6

**16** [The index tells us the number of factors of the base number that are multiplied together.]
$3^2$ means that two factors of 3 are multiplied together.
$3^2 = 3 \times 3$

## Standard level questions

**Mini Test 8:** Mixed Questions .................. Page 8

**1** C **2** C **3** C **4** \$564.20 **5** A **6** 194 **7** B **8** \$168 **9** \$0.85 **10** A **11** 46 **12** A **13** D **14** 310 **15** 46 **16** B

**1** Fill in the positions on the number line:

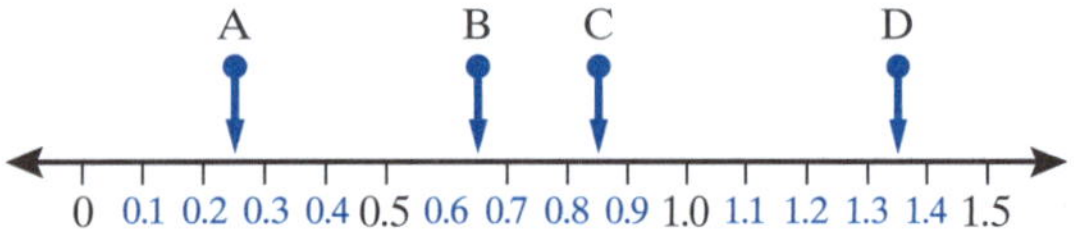


0.85 is between 0.8 and 0.9.
The arrow that points to 0.85 is C.

**2** A pentagon has 5 sides.
The only shape with 5 sides is shape C.

**3** The sector for Joe is about one-eighth of the circle. This means that about one-eighth of the people voted for Joe.

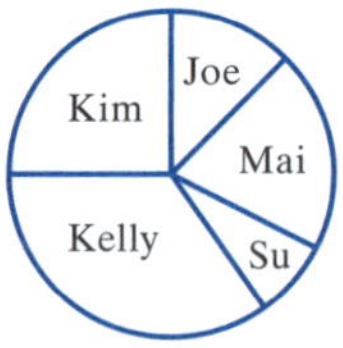


Number who voted for Joe = 800 ÷ 8
= 100
About 100 people voted for Joe.

**4** Cost of timber = 15.5 × \$36.40
= \$564.20

**5** There are more 1s than any other number on the spinner so the arrow is most likely to stop on 1.

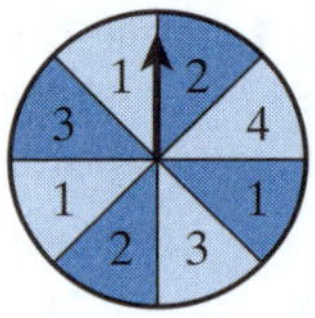


**6** [A total of \$582 was paid. \$3 was paid for each goat so you need to know how many lots of \$3 there are in \$582.]
Number of goats = \$582 ÷ \$3
= 194
194 goats were tested.

**7** A cube has 6 faces, 12 edges and 8 vertices.
Total = 6 + 12 + 8
= 26

**8** Total raised = 24 × \$7
= \$168

**9** Cost per can = \$27.20 ÷ 32
= \$0.85

**10** From 4:38 pm until 5 pm is 22 minutes.
From 5 pm until 5:17 pm is 17 minutes.
Total time = (22 + 17) min
= 39 min

**11** Ajak's object has 18 blocks now.

When finished it will be 4 blocks high, so it must also be 4 blocks long and 4 blocks wide (because it is a cube).

Total blocks in cube = 4 × 4 × 4
= 64
Extra blocks needed = 64 − 18
= 46

**12** The perimeter is 144 metres.
A square has 4 sides of equal length.
Length of each side = 144 m ÷ 4
= 36 m

**13** Angles on a straight line add to 180°.

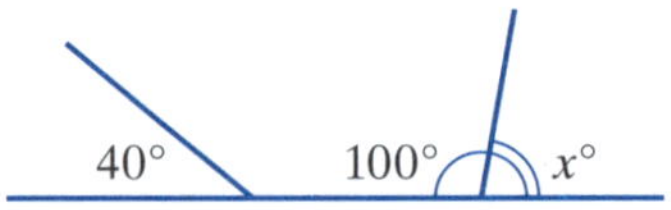


So 100 + $x$ = 180
$x$ = 180 − 100
= 80
[The angle of 40° has no effect on the value of $x$.]

**14** 700, 622, 544, 466, 388

700 − 622 = 78
622 − 544 = 78
544 − 466 = 78
466 − 388 = 78

The numbers are decreasing by 78 each time.

Next number = 388 − 78
= 310

**15** Number of spectators surveyed = 180

| | Birds | Serpents |
|---|---|---|
| Males | 38 | 69 |
| Females | 27 | |

Number in table = 38 + 27 + 69
= 134

Difference = 180 − 134
= 46

So the number of females who expected the Serpents to win is 46.

**16** Area $= \frac{1}{2} \times \text{base} \times \text{height}$

$= \frac{1}{2} \times 12\text{ m} \times 8\text{ m}$

$= 48\text{ m}^2$

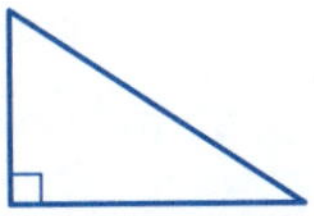


**Intermediate level questions**

## Mini Test 9: Fractions, Decimals and Percentages ........ Page 9

**1** 36 **2** $40 000 **3** 0.44 **4** 0.46 **5** C **6** B **7** D **8** A **9** D **10** 0.6003, 0.603, 0.6033, 0.63 **11** 0.6 **12** C **13** C **14** D **15** 2.75 kg **16** B **17** $720 **18** B **19** D **20** A **21** C **22** A **23** 24 **24** 63.58

**1** 15% of 240

10% of 240 $= \frac{1}{10}$ of 240
= 24

Now 5% $= \frac{1}{2}$ of 10%

So 5% of 240 = 12
15% = 10% + 5%
So 15% of 240 = 24 + 12
= 36

**2** 30% is $12 000.
So 10% is $4000 [$12 000 ÷ 3]
and 100% is $40 000. [$4000 × 10]
The total collection was $40 000.

**3** [The number halfway between two numbers is the average of those two numbers.]

Average $= \frac{0.35 + 0.53}{2}$

$= \frac{0.88}{2}$

= 0.44

The number halfway between 0.35 and 0.53 is 0.44.

**4** 0.7, 0.64, 0.58, 0.52

0.7 = 0.70
0.70 − 0.64 = 0.06

$$\begin{array}{r} 0.{}^{6}\not{7}{}^{1}0 \\ -\ 0.6\ 4 \\ \hline 0.0\ 6 \end{array}$$

0.64 − 0.58 = 0.06
0.58 − 0.52 = 0.06

The pattern is that the numbers are decreasing by 0.06 each time.

Next number = 0.52 − 0.06
= 0.46

$$\begin{array}{r} 0.{}^{4}\not{5}{}^{1}2 \\ -\ 0.0\ 6 \\ \hline 0.4\ 6 \end{array}$$

**5** $\frac{2}{3} + \frac{3}{5} = \frac{10}{15} + \frac{9}{15}$

$= \frac{19}{15}$

$= 1\frac{4}{15}$

**6** Increase = $300 − $240
= $60

% increase $= \frac{\$60}{\$240} \times 100\%$

$= \frac{1}{4} \times 100\%$

= 25%

**7** $\frac{2}{3}$ of the number = 12

$\frac{1}{3}$ of the number = 12 ÷ 2
= 6

The number = 6 × 3
= 18

**8** $11 + 1.1 + 0.11 = 11.00 + 1.10 + 0.11$
$= 12.21$

$$\begin{array}{r} 11.00 \\ 1.10 \\ +\ 0.11 \\ \hline 12.21 \end{array}$$

**9** [To divide by a fraction, multiply by the reciprocal.]

$8 \div \frac{1}{2} = 8 \times 2$
$= 16$

[The question is asking 'How many halves are in 8?' If 8 is divided into halves, there will be 16 halves altogether.]

**10** 0.63 0.6003 0.6033 0.603

[Write each decimal with the same number of places after the decimal point.]

0.6300
0.6003
0.6033
0.6030

In order from lowest to highest the numbers are: 0.6003, 0.6030, 0.6033, 0.6300 or 0.6003, 0.603, 0.6033, 0.63.

**11** $0.2 + 0.7 + 0.9 = 1.8$

$$\begin{array}{r} 0.2 \\ 0.7 \\ +\ 0.9 \\ \hline 1.8 \end{array}$$

Average $= 1.8 \div 3$
$= 0.6$

$$\begin{array}{r} 0.\ 6 \\ 3\overline{)1.{}^{1}8} \end{array}$$

**12** [You want each number to have only one digit after the decimal point. So you need to look at the digit in the second place after the decimal point to see whether you need to round up.]

0.576 = 0.6 correct to one decimal place

[The second digit is 7 which is greater than 5, so you need to round up.]

4.6183 = 4.6 correct to one decimal place

[The second digit is 1, which is less than 5, so you do not need to round up.]

The best approximations for 0.576 and 4.6183 are 0.6 and 4.6.

**13** $1\frac{1}{2} \times 1\frac{1}{3} = \frac{3}{2} \times \frac{4}{3}$
$= \frac{12}{6}$
$= 2$

[To multiply fractions first convert them to improper fractions. Then multiply the numerators and multiply the denominators, or use cancelling: $\frac{{}^{1}\cancel{3}}{{}_{1}\cancel{2}} \times \frac{{}^{2}\cancel{4}}{{}_{1}\cancel{3}}$]

**14** $\frac{1}{2} = 0.5$

$50\% = 0.5$

The three numbers are all equal.

**15** Amount of flour
$= (2.5 - 0.25 + 0.5)$ kg
$= 2.75$ kg

$$\begin{array}{r} 2.{}^{4}\cancel{5}{}^{1}0 \\ -\ 0.2\ 5 \\ \hline 2.2\ 5 \end{array} \qquad \begin{array}{r} 2.25 \\ +\ 0.50 \\ \hline 2.75 \end{array}$$

**16** $3 - 1\frac{3}{4} = 2 - \frac{3}{4}$
$= 1\frac{1}{4}$

[Or change to improper fractions with the same denominator. $\frac{12}{4} - \frac{7}{4} = \frac{5}{4} = 1\frac{1}{4}$]

**17** Precious saved 20% of \$900.

Now 10% of \$900 is \$90.

So 20% of \$900 is 2 × \$90 or \$180.

So Precious saved \$180.

Amount she paid = \$900 − \$180
= \$720

**18** Consider each option:

$\frac{2+4}{5+4} = \frac{6}{9} = \frac{2}{3} \neq \frac{2}{5}$

$\frac{2 \times 4}{5 \times 4} = \frac{8}{20} = \frac{2}{5}$

$\frac{2^2}{5^2} = \frac{4}{25} \neq \frac{2}{5}$

$\frac{10 \div 2}{10 \div 5} = \frac{5}{2} \neq \frac{2}{5}$

The fraction that is equal to $\frac{2}{5}$ is $\frac{2 \times 4}{5 \times 4}$.

**19** $6 \div 0.3 = 60 \div 3$
$= 20$

[The question could also be done by estimation. It is asking how many 0.3s are in 6. There are just over 3 of them in 1, so there will be a little more than 18 (6 × 3) of them in 6. Of the options, the only possible choice is 20.]

**20** $\frac{3}{10} + \frac{17}{100} = \frac{30}{100} + \frac{17}{100}$
$= \frac{47}{100}$
$= 0.47$

[Or change both fractions to decimals first: $0.3 + 0.17 = 0.47$]

**21** $\frac{2}{3}, \frac{7}{12}, \frac{5}{6}, \frac{3}{4}$
The common denominator is 12.
$\frac{2}{3} = \frac{8}{12}$ (after multiplying both numerator and denominator by 4)
$\frac{5}{6} = \frac{10}{12}$ (after multiplying both numerator and denominator by 2)
$\frac{3}{4} = \frac{9}{12}$ (after multiplying both numerator and denominator by 3)
So, in order, the fractions are $\frac{7}{12}, \frac{8}{12}, \frac{9}{12}$ and $\frac{10}{12}$.
The largest number is $\frac{10}{12}$ or $\frac{5}{6}$.

**22** $\frac{12+6}{2 \times 3} = \frac{18}{6}$
$= 3$

**23** 70% are male so 30% (100% − 70%) are female.
Now 10% of 80 = 8.
So 30% of 80 = 3 × 8
= 24
The number of female workers is 24.

**24** 187 × 34 = 6358
18.7 × 3.4 = 63.58
[There are 2 digits after decimal points in the question so there must be 2 digits after the decimal point in the answer.]

## Intermediate level questions

**Mini Test 10:** Angles and Measurements.................................................... Page 10

**1** 112° **2** A **3** 12 **4** C **5** 70 **6** 645 **7** A, F **8** C **9** C **10** 115 **11** B **12** 40 **13** 30 **14** 150 **15** A **16** 5000 **17** 0.003 km, 7 m, 20 000 mm, 4000 cm

**1** The size of the angle is 112°.
[The angle is measured from the right-hand side so use the scale that begins at 0° on the right; the inside scale. ]
[The angle is 2 more degrees than 110°, so is 112°.]

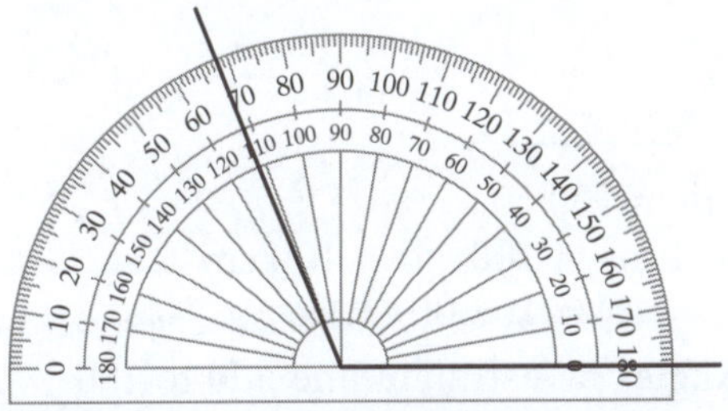

**2** A reflex angle measures more than 180° but less than 360°.
The only angle that is more than 180° is A.

[B is obtuse, C is a right angle and D is acute.]

**3** 3 L = 3000 mL
Number of glasses = 3000 ÷ 250
= 12

**4** Angles that measure between 90° and 180° are obtuse, so an angle of 110° is obtuse.
[Acute angles measure less than 90° and reflex angles measure between 180° and 360°. Two angles are complementary if they add to 90°.]

**5** The diagram shows a right angle divided into two parts.

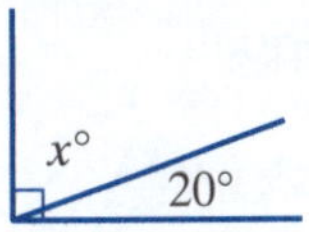


A right angle measures 90°.
So $x + 20 = 90$
This means that $x = 70$.

**6** 10 hours = 10 × 60 minutes
= 600 minutes
$\frac{3}{4}$ hour = 45 minutes
So $10\frac{3}{4}$ hours = 645 minutes

**7** 1 t = 1000 kg
1 tonne and 200 kilograms = (1000 + 200) kg
= 1200 kg
= 1.2 t
So both 1.2 t and 1200 kg are the same as 1 tonne and 200 kilograms.

**8** The angle is larger than a right angle but less than a straight angle. So the angle is between 90° and 180°. The angle is closer in size to a right angle than a straight angle.

The best estimate for the size of the angle is 120°.

**9** 1 m = 100 cm
So 1 $m^2$ = 100 cm × 100 cm
= 10 000 $cm^2$

**10** Angles in a straight line add to 180°.
So $x + 65 = 180$
$x = 180 - 65$
$= 115$

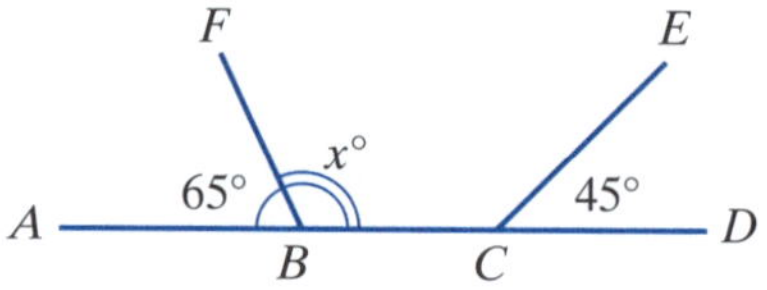


[The angle of 45° has no effect on the angle marked $x$°.]

**11** An angle of 90° is a right angle. An angle of 50° is a little more than half the size of a right angle. So the angle that is about 50° is B.

[A would be about 70°, C about 20° and D about 110°.]

**12** Vertically opposite angles are equal.
So shaded angle + 90° = 130°
Shaded angle = 130° − 90°
= 40°

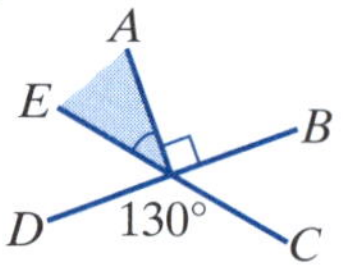


**13** Angles at a point add to 360°.
So $x + 110 + 90 + 130 = 360$
$x + 330 = 360$
$x = 30$

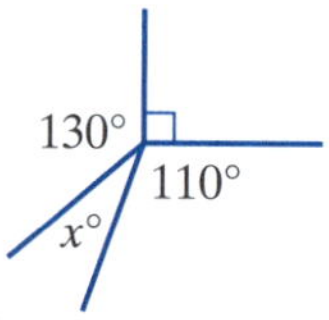


**14** Each angle of a square is 90°.
Each angle of an equilateral triangle is 60°.
So $x = 90 + 60$
$= 150$

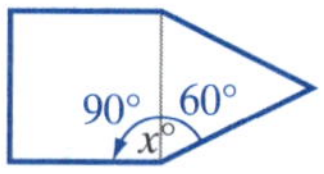


**15** If an isosceles triangle is cut along an axis of symmetry the two triangles formed must be right-angled.
[The two triangles will be isosceles if the original triangle is a right-angled triangle. Otherwise they will be scalene. (So they could be either isosceles or scalene, but they must be right-angled.)]

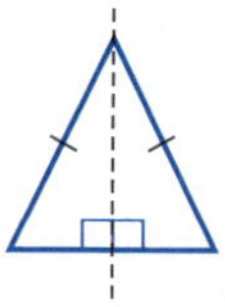

**16** 1 ML = 1 000 000 L
1 kL = 1000 L
5 ML = 5 000 000 L
= 5000 kL

**17** 1 km = 1000 m
So 0.003 km = 0.003 × 1000 m
= 3 m
100 cm = 1 m
So 4000 cm = 40 m
1000 mm = 1 m
So 20 000 mm = 20 m
So, from shortest to longest, the lengths are 3 m, 7 m, 20 m and 40 m or 0.003 km, 7 m, 20 000 mm and 4000 cm.

## Intermediate level questions

**Mini Test 11:** Chance and Data ............... Page 11

**1** C **2** B **3** 48 **4** B **5** D **6** 87 **7** C **8** 4 **9** D
**10** C **11** 15

**1** 12 of the 30 pegs are red.

Chance of red $= \frac{12}{30}$

$= \frac{2}{5}$ [after dividing both numerator and denominator by 6]

**2** There are 10 scores.
Sum of scores = 4 × 7 + 3 × 8 + 2 × 9 + 10
= 28 + 24 + 18 + 10
= 80
Average = 80 ÷ 10
= 8
Now the average does not change when a score is left out. So the score that is left out must be the same as the average.
The score that is left out is 8.

**3** Of the 150 questioned, 71 were girls.
Number of boys = 150 − 71
= 79
Of the 79 boys, 31 had seen the movie.
Number of boys who hadn't seen the movie
= 79 − 31
= 48

| | Seen | Not seen | Total |
|---|---|---|---|
| Boys | 31 | **48** | **79** |
| Girls | 17 | 54 | 71 |
| Total | 48 | **102** | 150 |

**4** Total marbles = 11 + 17 + 9 + 8
= 45
9 out of 45 marbles are yellow.

Chance of yellow marble $= \frac{9}{45}$
$= \frac{1}{5}$

**5**

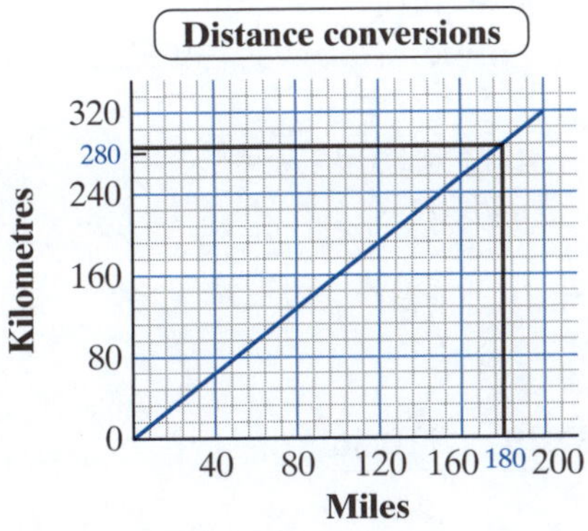


180 miles is about 280 km.

**6** Total from 5 tests = 5 × 81
= 405
Total from 6 tests = 6 × 82
= 492
Difference = 492 − 405
= 87
Henning would have scored 87 in his sixth test.

**7** Dimple could get 1 on the first die and 4 on the second, 2 on the first and 3 on the second, 3 on the first and 2 on the second, or 4 on the first die and 1 on the second die. There are 4 ways in which Dimple could score 5.

**8** Time on Tuesday $= 1\frac{1}{4}$ h
= 1 h 15 min
= 75 min
Time on Thursday $= 1\frac{1}{2}$ h
= 1 h 30 min
= 90 min
Time on Friday = 1 h 10 min
= 70 min
Total time = (45 + 75 + 40 + 90 + 70) min
= 320 min
Average time = 320 min ÷ 5
= 64 min
= 1 h 4 min
The number that should be placed in the box is 4.

**9** The spinner has 12 sections. In 12 spins you would expect it to land on each section once. So in 12 spins you would expect to land on red once, green twice, yellow 4 times and blue 5 times.
In 24 spins you would expect to land on red twice, green 4 times, yellow 8 times and blue 10 times.

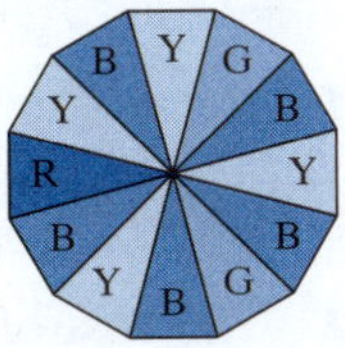


The colour that you would expect to get 8 times is yellow.

**10** Consider each option:
A: There are 7 balls and 1 is blue.
Chance of blue is 1 in 7.

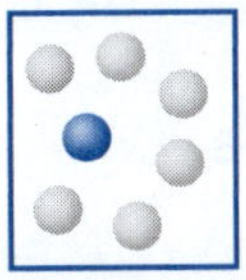

B: There are 5 balls and 1 is blue.
Chance of blue is 1 in 5.

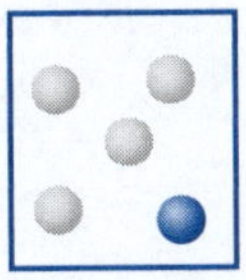

C: There are 12 balls and 2 are blue.
Chance of blue is 2 in 12 or 1 in 6.

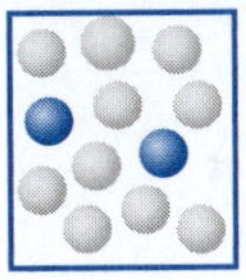

D: There are 12 balls and 3 are blue.
Chance of blue is 3 in 12 or 1 in 4.

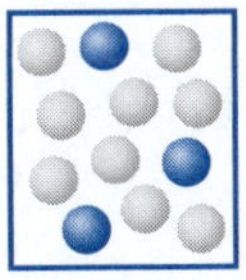

The option where the chance is 1 in 6 is C.

**11**

| Mark | 5 | 6 | 7 | 8 | 9 | 10 |
|---|---|---|---|---|---|---|
| Number | 8 | 12 | 17 | 15 | 11 | 9 |

Number scoring more than 7
$= 15 + 11 + 9$
$= 35$
Number scoring less than 7 $= 8 + 12$
$= 20$
Difference $= 35 - 20$
$= 15$
So 15 more students scored more than 7 than scored less than 7.

## Intermediate level questions

**Mini Test 12:** Number Page 12

**1** D **2** \$8.10 **3** A **4** D **5** 0.75 kg **6** −1.5 **7** B, E **8** A **9** D **10** 2.92 **11** 1382 **12** 5 **13** 8 and 3 **14** B **15** B **16** C **17** B **18** C **19** C **20** A **21** \$31.30 **22** 4.12 **23** 1440 mL **24** B

**1** $10^2 - 6^2 = 100 - 36$
$= 64$
$= 8^2$

**2** The cost of 8 cans = \$21.60
The cost of 1 can = \$21.60 ÷ 8
= \$2.70
The cost of 3 cans = 3 × \$2.70
= \$8.10

**3** $39 \times 27 - 19 \times 39 = 39 \times 27 - 39 \times 19$
$= 39 \times (27 - 19)$
$= 39 \times 8$

**4** $\frac{3}{7} = 3 \div 7$
$= 0.428571428\ldots$
$= 0.43$ (correct to 2 decimal places)
[The third digit after the decimal place is 8 which is greater than 5 so the decimal is rounded up.]

**5** [Divide the total cost by the cost per kilogram and the answer will be the number of kilograms.]
Number of kilograms = \$6.30 ÷ \$8.40
= 0.75

**6** The arrow is pointing to −1.5. [or $-1\frac{1}{2}$]

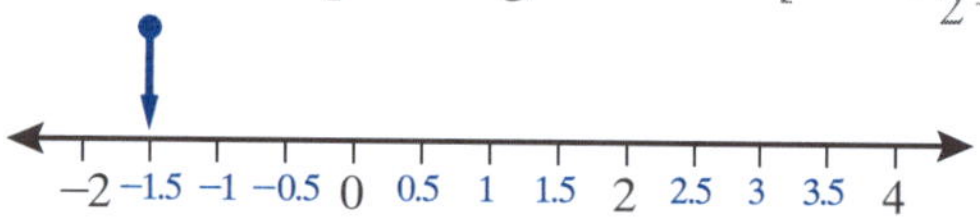


**7** [A prime number has exactly 2 factors; 1 and itself.]
55 ends in 5 so it is divisible by 5. It is not prime.
The sums of the digits of 51 and 57 are divisible by 3 ($5 + 1 = 6, 5 + 7 = 12$), so 51 and 57 are both divisible by 3. They are not prime.
The numbers that are prime are 53 and 59.

**8** After multiplying by 7, Shelly's answer was 490.
Before multiplying by 7 it must have been $490 \div 7 = 70$.
Now 70 should have been divided by 7.
The answer should have been $70 \div 7 = 10$.

**9** The temperature was 2 degrees below zero at 6 pm. At 6 am it was another 7 degrees below that.
Now $2 + 7 = 9$, so the temperature was 9 degrees below zero.
At 6 am the temperature was −9°C.

**10** $3.26 + 4.83 = 5.17 + \square$
Now $3.26 + 4.83 = 8.09$
So $5.17 + \square = 8.09$
$8.09 - 5.17 = 2.92$
So $5.17 + 2.92 = 8.09$
$\square = 2.92$

**11** $\frac{147.85 - 37.29}{0.08} = \frac{110.56}{0.08}$
$= 1382$

**12** After 17 was subtracted, the answer was 8.
Before 17 was subtracted, the answer was 8 + 17 or 25.
The positive number that multiplies by itself to give 25 is 5.
The number was 5.

**13** $8 \times 3 = 24$
$8 + 3 = 11$
The numbers must be 8 and 3.
[These numbers are found by trial and error. For example, $6 \times 4 = 24$ but $6 + 4 = 10$, not 11, so the numbers are not 6 and 4. (The answer can be given in either order: 3 and 8 or 8 and 3.)]

**14** 16 out of 24 chocolates have cream centres.
Fraction of cream centres
$= \frac{16}{24}$
$= \frac{2}{3}$ [after dividing both numerator and denominator by 8]

**15** The cost of the scooter ($77) is closer to $80 than $70.
The cost of the doll ($28) is closer to $30 than $20.
The cost of the clothes ($132) is closer to $130 than $140.
The best way to estimate the cost is $80 + $30 + $130.

**16** $2^3 \times 4^2 = 2 \times 2 \times 2 \times 4 \times 4$
[But this is not one of the options.]
Now $2 \times 2 = 4$
So $2 \times 2 \times 2 \times 4 \times 4 = 2 \times (2 \times 2) \times 4 \times 4$
$= 2 \times 4 \times 4 \times 4$

**17** Multiples of 6 are 6, 12, 18, 24, 30, …
Multiples of 8 are 8, 16, 24, 32, 40, …
24 is a multiple of both 6 and 8.
So 24 has both 6 and 8 as factors.
[Or simply divide each of the options by 6 and by 8 to see into which number they both divide evenly.]

**18** $-2 + 5 \times 2 \square 6 - 3 \times 2$
$-2 + 5 \times 2 = -2 + 10$
$= 8$
$6 - 3 \times 2 = 6 - 6$
$= 0$
Now 8 is greater than 0.
So $-2 + 5 \times 2$ is greater than $6 - 3 \times 2$.
The correct symbol is $>$.

**19** $\frac{2}{3}$ of a number is 24.
$\frac{1}{3}$ of the number $= 24 \div 2$
$= 12$
The number $= 12 \times 3$
$= 36$

**20** $\sqrt{25 + 144} = \sqrt{169}$
$= 13$

**21** Alice could buy 5 packets of 3 batteries.
Cost $= 5 \times \$6.40$
$= \$32$
Alice could also buy 3 packets of 4 batteries and 1 packet of 3 batteries.
Cost $= 3 \times \$8.30 + \$6.40$
$= \$24.90 + \$6.40$
$= \$31.30$
The least amount Alice could pay is $31.30.

**22** [The number halfway between two others is the average of the two other numbers.]
$2.78 + 5.46 = 8.24$
$8.24 \div 2 = 4.12$
The number halfway between 2.78 and 5.46 is 4.12.

**23** $\frac{1}{6}$ of the capacity is 240 mL.
The capacity $= 6 \times 240$ mL
$= 1440$ mL

**24** Decrease $= 200 - 120$
$= 80$
So the decrease is 80 out of 200.
This is a decrease of 40 out of 100.
The percentage decrease is 40%.

## Intermediate level questions

**Mini Test 13:** Problems ................................ Page 13

**1** $4.75 **2** B **3** 13 **4** 6 **5** $126 720
**6** English, Science, Maths, Art **7** C **8** $0.80
**9** $749 **10** $900 **11** 27 **12** 23 **13** 12 **14** −6°C
**15** D **16** $2.30

**1** Cost of 1 can $= \$8.55 \div 9$
$= \$0.95$
Cost of 5 cans $= 5 \times \$0.95$
$= \$4.75$

**2** 3 balloons were left over.
Balloons shared $= 45 - 3$
$= 42$
Each person sharing had 6 balloons.
Number of people sharing $= 42 \div 6$
$= 7$
Nikita is one of those 7 people.
Number of Nikita's friends $= 7 - 1$
$= 6$
Nikita shared the balloons with 6 friends.

**3** After multiplying by 7 the answer was 133.
Before multiplying by 7, the result must have been $133 \div 7$ or 19.
Before adding 6, the number must have been $19 - 6$ or 13.
The number was 13.

# Year 7 Numeracy Mini Test Answers

**4** When each of the 8 bags holds 5 marbles there are 2 left over. If those 2 marbles are placed in two bags, those bags will have 6 marbles. There are another 6 bags that need an extra marble so 6 more marbles will be needed.

[Or, the number of marbles = $5 \times 8 + 2$
= 42

The number of marbles needed = $6 \times 8$
= 48

Extra marbles needed = $48 - 42$
= 6 ]

**5** Monthly rent = \$1320

Monthly income = $8 \times$ \$1320
= \$10 560

Yearly income = $12 \times$ \$10 560
= \$126 720

**6** Change each result to a mark out of 100.

English: $\frac{81}{100}$

Maths: $\frac{39}{50} = \frac{78}{100}$

Science: $\frac{32}{40} = \frac{8}{10} = \frac{80}{100}$

Art: $\frac{45}{60} = \frac{15}{20} = \frac{75}{100}$

In order, from the highest percentage mark to the lowest, the subjects are English, Science, Maths and Art.

**7** [Find the cost of 1 kg for each option.]

4 kg for \$4.20 = 1 kg for \$1.05

5 kg for \$4.90 = 1 kg for \$0.98

6 kg for \$5.70 = 1 kg for \$0.95

3 kg for \$3.15 = 1 kg for \$1.05

The cheapest price per kilogram is \$0.95.

The cheapest onions are 6 kg for \$5.70.

**8** The pen and ruler cost \$2.90.

If the pen cost \$1.30 less, it would then cost the same as the ruler.

Now \$2.90 − \$1.30 = \$1.60

Cost of the ruler = \$1.60 ÷ 2
= \$0.80

The ruler cost 80 cents (and the pen cost \$2.10).

**9** Amount spent on food = \$1605 ÷ 3
= \$535

Amount spent on rent = \$1605 ÷ 5
= \$321

Total = \$535 + \$321
= \$856

Remainder = \$1605 − \$856
= \$749

Sarah would have \$749 left to spend on other things.

**10** Cash price = \$9500

Cost to Lara = \$2000 + 12 × \$700
= \$2000 + \$8400
= \$10 400

Difference = \$10 400 − \$9500
= \$900

Lara will pay an extra \$900.

**11** Jerome has 36 cars and trucks. He has 3 times as many trucks and cars so one-quarter are cars and three-quarters are trucks.

Number of cars = $36 \div 4$
= 9

Number of trucks = $9 \times 3$
= 27

Jerome has 27 trucks.

**12** Luke is 2 years older than Jim and 5 years older than Pete. So, if Jim was 2 years older and if Pete was 5 years older they would both be the same age as Luke.

Now $62 + 2 + 5 = 69$

So if they were all the same age as Luke the sum of their ages would be 69.

Now $69 \div 3 = 23$

So they would all be 23.

Luke is 23 years old.

**13** Number of plants = $8 \times 18$
= 144

In 12 rows, the number per row = $144 \div 12$
= 12

There would be 12 plants in each row.

**14** The maximum temperature on Monday was 15°. If the temperature was 15° colder it would be 0°.

But the temperature was 21° colder.

Now $21 - 15 = 6$

So the temperature fell another 6°. This means it was 6° below zero.

The minimum temperature on Tuesday was −6 °C.

**15** For every 2 adults there are 5 children. So 5 out of every 7 people at the fete are children.

Now $140 \div 7 = 20$

So there are 20 lots of 7 people at the fete.

Number of children = $5 \times 20$
= 100

There are 100 children at the fete (and 40 adults).

# Year 7 Numeracy Mini Test Answers

**16** Cost of 5 return tickets $= 5 \times \$2.70$
$= \$13.50$
Saving $= \$13.50 - \$11.20$
$= \$2.30$

## Intermediate level questions

**Mini Test 14:** Patterns ................................ Page 14

**1** 39 **2** C **3** D **4** 60 **5** 15 **6** 25 **7** B **8** A **9** 48
**10** C **11** 63 **12** A **13** 35 **14** B **15** 17

**1** 3, 7, 11, 15
The numbers are increasing by 4 each time.
Continue the pattern:
3, 7, 11, 15, 19, 23, 27, 31, 35, 39, 43, …
The 10th number is 39.

**2**

| $\triangle$ | 2 | 3 | 4 | 5 | 6 |
|---|---|---|---|---|---|
| $\square$ | 4 | 7 | 10 | 13 | 16 |

Consider each option:
$\square = 2 \times \triangle$
When $\triangle = 2$,
$\square = 2 \times 2$
$= 4$ ✓
When $\triangle = 3$,
$\square = 2 \times 3$
$= 6$ ✗
This is not the option.
$\square = \triangle + 2$
When $\triangle = 2$,
$\square = 2 + 2$
$= 4$ ✓
When $\triangle = 3$,
$\square = 3 + 2$
$= 5$ ✗
This is not the option.
$\square = 3 \times \triangle - 2$
When $\triangle = 2$,
$\square = 3 \times 2 - 2$
$= 6 - 2$
$= 4$ ✓
When $\triangle = 3$,
$\square = 3 \times 3 - 2$
$= 9 - 2$
$= 7$ ✓
When $\triangle = 4$,
$\square = 3 \times 4 - 2$
$= 12 - 2$
$= 10$ ✓
When $\triangle = 5$,
$\square = 3 \times 5 - 2$
$= 15 - 2$
$= 13$ ✓
When $\triangle = 6$,
$\square = 3 \times 6 - 2$
$= 18 - 2$
$= 16$ ✓
This is the option.
The correct rule is $\square = 3 \times \triangle - 2$.

**3** $111 - 7 = 104$
$104 - 7 = 97$
$97 - 7 = 90$
The numbers are 111, 104, 97 and 90.

**4** Write out the terms:
6, 15, 24, 33, 42, 51, 60, 69, 78, ….
The seventh number is 60.

**5** 1, 3, 6, 10, ?, 21, 28, 36
Find the differences between successive terms:
$3 - 1 = 2$
$6 - 3 = 3$
$10 - 6 = 4$
The differences are increasing by 1 each time.
The next difference will be 5.
$10 + 5 = 15$.
The missing number is 15.
[Check $15 + 6 = 21$, $21 + 7 = 28$, …]

**6**

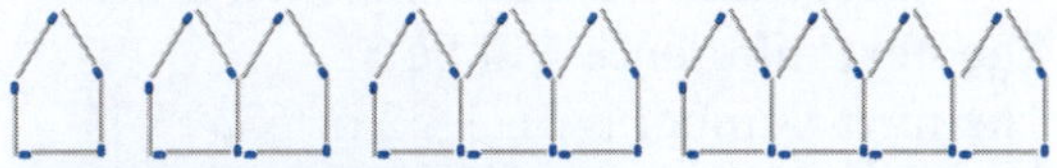

| Number of pentagons | 1 | 2 | 3 | 4 |
|---|---|---|---|---|
| Number of matchsticks | 5 | 9 | 13 | 17 |

The pattern for the number of matchsticks is 5, 9, 13, 17, …
The numbers are increasing by 4 each time.
Continue the pattern:
5, 9, 13, 17, 21, 25, 29, …
The sixth number is 25 so 25 matchsticks would be needed to make 6 pentagons.

**7** [Study the pairs of numbers to find the rule.]

| First number | 1 | 2 | 3 | 4 | 5 |
|---|---|---|---|---|---|
| Second number | 5 | 11 | ? | 23 | 29 |

The rule is to multiply the first number by 6 and subtract 1.
So when the first number is 3, the second number is $6 \times 3 - 1$ or 17.

**8** $\frac{1}{5}, \frac{1}{4}, \frac{3}{10}, \frac{7}{20}, \ldots$
Change the fractions to equivalent fractions with a common denominator of 20.
$\frac{4}{20}, \frac{5}{20}, \frac{6}{20}, \frac{7}{20}, \ldots$
So the next number will be $\frac{8}{20}$.
Now $\frac{8}{20} = \frac{2}{5}$.
The next number is $\frac{2}{5}$.

**9** 3, 6, 12, 24, …
The numbers are doubling each time.
The next number will be double 24 or 48.

**10**

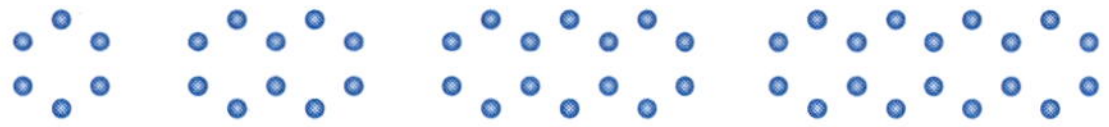

| Number of hexagons | 1 | 2 | 3 | 4 |
|---|---|---|---|---|
| Number of pins | 6 | 10 | 14 | 18 |

The pattern for the number of pins is 6, 10, 14, 18, …
The numbers are increasing by 4 each time.
Continue the pattern:
6, 10, 14, 18, 22, 26, 30, 34, …
30 is the seventh number in the pattern so 30 pins will be needed to make 7 hexagons.

**11** 1, 3, 7, 15, 31, …
The differences are 2, 4, 8 and 16.
The differences are doubling each time.
The next difference will be 32.
The next number is $31 + 32$ or 63.
[Or, to find the next number double the previous number and add 1. So the next number is $2 \times 31 + 1 = 62 + 1 = 63$.]

**12** C, B, A, D, C, B, A, D, C, B, A, D, C, B,…
There are 4 letters that are repeating. Every fourth letter will be D.
So the 44th letter will be D.
The 43rd letter will be the letter before D in the pattern.
The 43rd letter will be A.

**13** Write out the pattern:
3, 5, 7, 9, 11, 13, …
Sum of first 5 terms $= 3 + 5 + 7 + 9 + 11$
$= 35$

**14**

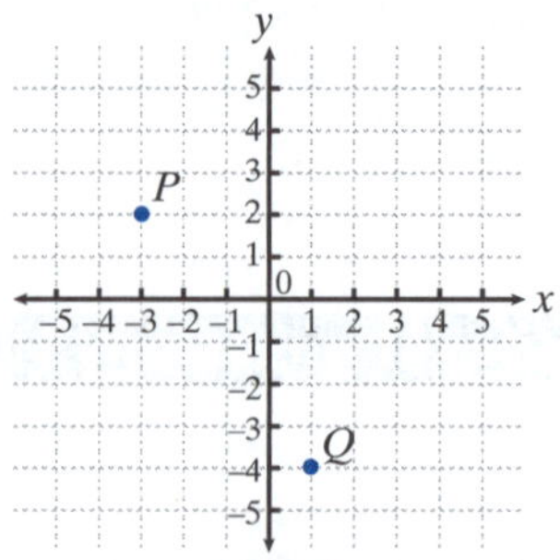


$P$ has an $x$-value of $-3$ and $y$-value of 2.
$P$ is the point $(-3, 2)$.
$Q$ has an $x$-value of 1 and $y$-value of $-4$.
$Q$ is the point $(1, -4)$.
$P$ is at $(-3, 2)$ and $Q$ at $(1, -4)$.

**15** The first number is 2.
The second number $= 2 \times 2 + 3$
$= 4 + 3$
$= 7$
The third number $= 2 \times 7 + 3$
$= 14 + 3$
$= 17$

## Intermediate level questions

### Mini Test 15: Time ........ Page 15

**1** A **2** A **3** 3:30 pm **4** C **5** 12:16 pm **6** 40 min **7** C **8** D **9** 50 min **10** B **11** B **12** C **13** B **14** B **15** C **16** 9:07 **17** C

**1** From 9:48 am until 10:00 am is 12 minutes. From 10:00 am until 10:12 am is another 12 minutes. So from 9:48 am until 10:12 am is 24 minutes.
From 10:12 am until 9:12 pm is 11 hours.
So from 9:48 am until 9:12 pm is 11 hours and 24 minutes.

**2** 20:45 is a time in 24-hour time format.
[Subtract 12 from the hours, to find how many hours it is after 12 noon.]
Now $20 - 12 = 8$
So 20:45 is 8:45 pm.

**3** Auckland time is 4 hours ahead of Perth time so Perth time is 4 hours behind Auckland time. 4 hours before 7:30 pm is 3:30 pm.
The football match will be shown at 3:30 pm.

**4** When it is 9:30 am in Paris it is 7:30 pm in Hobart. From 9:30 am until 7:30 pm is 10 hours, so Hobart time is 10 hours ahead of Paris time.

# Year 7 Numeracy Mini Test Answers

When it is 6:00 pm Saturday in Paris it will be 10 hours later in Hobart. 10 hours after 6:00 pm is 4:00 am the next day.
So, when it 6:00 pm on Saturday in Paris it will be 4:00 am on Sunday in Hobart.

**5** 1 hour after 10:46 am is 11:46 am. Half an hour (or 30 minutes) after that is 12:16 pm.

**6** A quarter to six is 15 minutes before 6 o'clock.
6:25 is 25 minutes after 6 o'clock.
Total time = (15 + 25) min
= 40 min

**7** You need to find the date 21 days before 13 September.
13 days before 13 September is 31 August.
Another 8 days before that is 23 August.

**8** The minute hand turns through 90° in one-quarter of an hour or 15 minutes.
Now 270 = 3 × 90
So the minute hand will turn through 270° in three-quarters of an hour or 45 minutes.
The clock shows 25 minutes before 7 o'clock.
In 45 minutes' time it will be 20 minutes after 7 o'clock or 7:20.

**9** The clock shows 25 minutes after 4 o'clock or 4:25. From 4:25 until 5:00 is 35 minutes. 5:15 is 15 minutes past 5:00.
Total time = (35 + 15) min
= 50 min

**10** From 8:40 until 9:00 is 20 minutes.
From 9:00 until 10:00 is 1 hour.
From 10:00 until 10:15 is 15 minutes.
Total time = 1 h 35 min

**11** Beats in a minute = 80
Beats in an hour = 80 × 60
= 4800
Beats in a day = 4800 × 24
= 115 200
Beats in a week = 115 200 × 7
= 806 400
The heart will beat about 800 000 times in a week.

**12** 4 hours after 9:40 am is 1:40 pm.
$\frac{3}{4}$ of an hour = 45 min
20 minutes after 1:40 pm is 2:00.
Extra minutes = 45 − 20
= 25
The time will be 2:25 pm.

**13** From 8:40 am until 9:00 am is 20 minutes.
From 9:00 am until 9:15 am is 15 minutes.
So from 8:40 am until 9:15 am is 35 minutes.
From 9:15 am until 3:15 pm is 6 hours.
The total time is 6 h 35 min.

**14** The train left Newcastle $2\frac{3}{4}$ hours before 14:25.
Two hours before 14:25 is 12:25.
$\frac{3}{4}$ h = 45 min
25 minutes before 12:25 is 12:00 and another 20 minutes before that is 11:40.

**15** Consider each option:

A The first clock shows a quarter to three or 2:45 and the bottom clock shows 3:45.
The times are not the same.

B The first clock shows 10 minutes to 2 o'clock or 1:50. The bottom clock shows 10:10. The times are not the same.

C The first clock shows 20 minutes to seven or 6:40. The bottom clock also shows 6:40.
The times are the same.

D The first clock shows 25 minutes past 4 or 4:25. The bottom clock shows 5:20. The times are not the same.
The pair of clocks that show the same time are those in Option C.

**16**

| | | | | |
|---|---|---|---|---|
| **Lincoln** | 8:14 | 8:56 | 9:27 | 10:06 |
| **Putney** | 8:28 | 9:15 | 9:44 | 10:20 |
| **Zetland** | 8:49 | 9:37 | 10:05 | 10:39 |
| **City** | 9:02 | 9:50 | 10:18 | 10:51 |

A quarter past ten is 10:15. To be in the city before 10:15, Vera will need to catch the train that arrives in the city at 9:50. This train leaves Putney at 9:15. Vera will need to leave home at least 8 minutes before this. The latest time that Vera should leave home is 9:07.

**17** 45 minutes is three-quarters of an hour.
1 h + 45 mins = (1 + 0.75) h
= 1.75 h

## Intermediate level questions

**Mini Test 16:** Measurement Page 16

**1** A **2** 7 **3** B **4** D **5** 750 mL **6** 156 km **7** A **8** D **9** A **10** B **11** D **12** B **13** C **14** C **15** B **16** 0.07 t, 70 500 g, 75 kg

# Year 7 Numeracy Mini Test Answers

**1** 1 kg = 1000 g
1 kg + 60 g = (1000 + 60) g
= 1060 g
= 1.06 kg

**2** 3 blocks balance 1 ball. So twice as many blocks will balance twice as many balls.
So 6 blocks balance 2 balls.
7 blocks would balance 2 balls and a block.

**3** Area = 800 m × 500 m
= 400 000 $m^2$
Now 1 ha = 10 000 $m^2$
Area in hectares = 400 000 ÷ 10 000
= 40
So the area is 40 ha.

**4** 1 cm = 10 mm
So 1 cm + 20 mm = 10 mm + 20 mm
= 30 mm

**5** 1 L = 1000 mL
Each litre on the jug is divided into 4.
Each division = 1000 mL ÷ 4
= 250 mL
So the jug holds 1250 mL.

If 500 mL is poured out, the amount of juice remaining
= (1250 − 500) mL
= 750 mL

**6** 1 cm represents 20 km.
So 7.8 cm would represent 7.8 × 20 km or 156 km.

**7** 1 m = 100 cm so 1.8 m = 180 cm
Each brick is 15 cm high.
Number of layers = 180 ÷ 15
= 12
Each layer has 12 bricks.
Number of bricks = 12 × 12
= 144

**8** There are 60 seconds in a minute.
So 15 seconds is one-quarter of a minute.
One-quarter = 0.25
So 1 min + 15 s = 1 min + 0.25 min
= 1.25 min

**9** 5 cm represents 25 m.
So 1 cm represents 5 m.

**10** Five 20c pieces make $1.
The number of 20c pieces in $15
= 15 × 5
= 75
So Kip has 75 coins, each of which has a diameter of 28 mm.
Length of the line = 75 × 28 mm
= 2100 mm
= 2.1 m

**11** 1 m = 100 cm
1 m + 15 cm = (100 + 15) cm
= 115 cm
So Option C is not correct.
Now 115 cm is 1.15 m.
So Options A and B are not correct.
1 cm = 10 mm
115 cm = 1150 mm
Option D is correct.

**12** The length of the photo was 12 cm and is now 54 cm.
Number of times larger = 54 ÷ 12
= 4.5
So the width of the photo will be 4.5 times bigger as well.
New width = 4.5 × 8 cm
= 36 cm

**13** One kilolitre = 1000 litres
So 8 kL = 8000 L
and 8.75 kL = 8750 L.

**14** Remove 1 ball from each side of the balance.

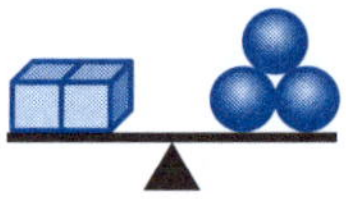

So 2 blocks balance 3 balls.
Twice as many blocks will balance twice as many balls.
So 4 blocks will balance 6 balls.

**15** From Aspen to Bunyip on the map is 9 squares.

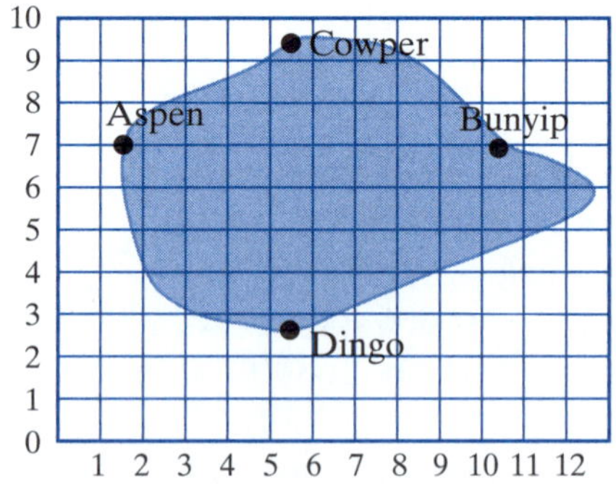


So 9 squares represent 6.75 km.
1 square would represent (6.75 ÷ 9) km or 0.75 km.
From Cowper to Dingo on the map is 7 squares.
7 squares represent 7 × 0.75 km or 5.25 km.

# Year 7 Numeracy Mini Test Answers

**16** 1 t = 1000 kg
so 0.07 t = 0.07 × 1000 kg
= 70 kg
1000 g = 1 kg
so 70 500 g = (70 500 ÷ 1000) kg
= 70.5 kg
In order from lightest to heaviest:

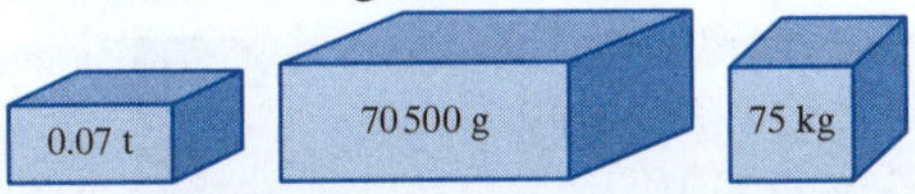


**Intermediate level questions**

**Mini Test 17:** Perimeter, Area and Volume ....... Page 17

**1** 150 cm$^2$ **2** 84 cm **3** B **4** 42 m **5** C **6** D
**7** 5 m **8** 25 m **9** B, D **10** 300 **11** 81 cm$^2$
**12** (see solution)

**1** Each face is a square of side 5 cm.

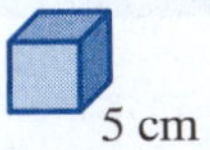


Area of each face = 5 cm × 5 cm
= 25 cm$^2$
There are 6 identical faces.
Total area = 6 × 25 cm$^2$
= 150 cm$^2$

**2** The top of the shape has 4 equilateral triangles. Two sides of each of these triangles make the border of the figure, so the top border is the same as 8 sides of a triangle. The right side is made up of 3 equilateral triangles and has 2 sides of each of these forming the border, so the right border is the same as 6 sides of a triangle.
The bottom is equal to 4 sides of a triangle and the left side is equal to 3 sides of the equilateral triangle.

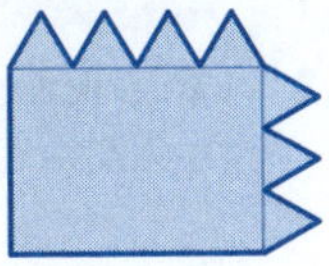

Total sides of a triangle = 8 + 6 + 4 + 3
= 21

Each side is 4 cm long.

Perimeter = 21 × 4 cm
= 84 cm

**3** Area of outer rectangle:
Area = 12 cm × 7 cm
= 84 cm$^2$

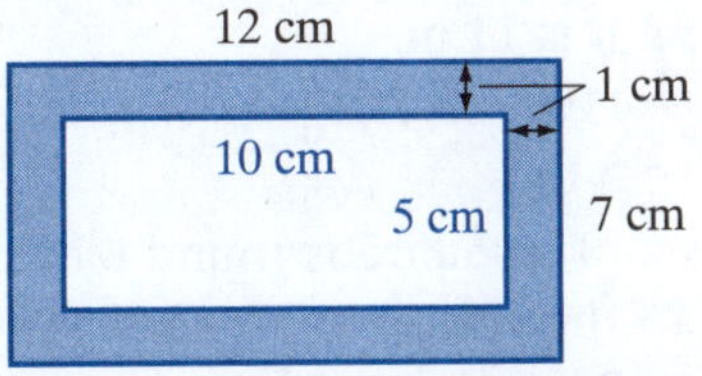


There is 1 cm on each side so the length of the inner rectangle is (12 − 2) cm or 10 cm. There is 1 cm at the top and 1 cm at the bottom so the width of the inner rectangle is (7 − 2) cm or 5 cm.
Area of inner rectangle:
Area = 10 cm × 5 cm
= 50 cm$^2$
Shaded area = (84 − 50) cm$^2$
= 34 cm$^2$

**4** The total of the top (horizontal) sides will be the same as the bottom side, 12 m.
The total of the right side = (2 + 2 + 5) m
= 9 m
The left side must also be 9 m.
Perimeter = 2 × (12 + 9) m
= 2 × 21 m
= 42 m

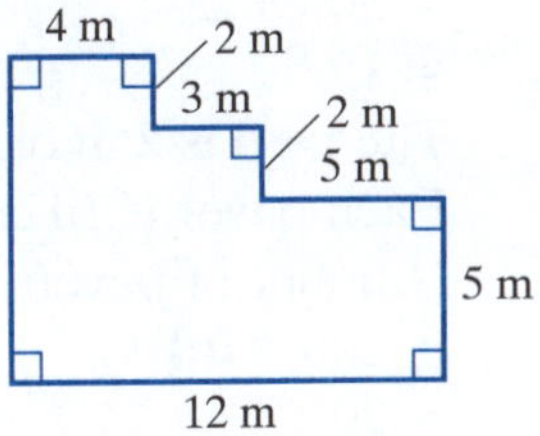


**5** If the second prism was twice as long but the same width and the same height it would have twice the volume.
Because it is also twice as wide as the first prism its volume is 4 times that of the first prism.
Volume = 4 × 12 m$^3$
= 48 m$^3$

**6** Paddock I:
Area = 60 m × 60 m
= 3600 m$^2$

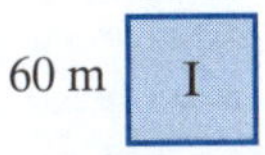


Paddock II:
Area = 120 m × 30 m
= 3600 m$^2$

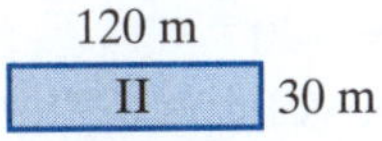


Paddock III:
Area = 90 m × 40 m
= 3600 m$^2$

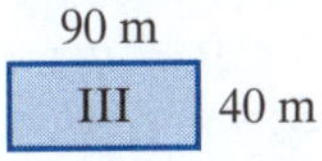


All three paddocks have the same area.

# Year 7 Numeracy Mini Test Answers

**7** The perimeter is 32 m.
The sum of the length and the width will be half of 32 m or 16 m.
The length is 11 m.
Width = (16 − 11) m
= 5 m

**8** The area of a square is found by squaring the length of the side.
Area = 625 $m^2$
Length of the side = $\sqrt{625}$ m
= 25 m

**9** 1 ha = 10 000 $m^2$ (not 100 $m^2$)
1 $km^2$ = 1000 m × 1000 m
= 1 000 000 $m^2$
= 100 ha
1 $m^2$ = 100 cm × 100 cm
= 10 000 $cm^2$
1 $cm^2$ = 10 mm × 10 mm
= 100 $mm^2$
The correct statements are 1 $km^2$ = 100 ha and 1 $cm^2$ = 100 $mm^2$.

**10** The yard is 3 m or 300 cm long.
Each paver is 20 cm long.
Number of pavers along the length
= 300 ÷ 20
= 15
The yard is 2 m or 200 cm wide.
Each paver is 10 cm wide.
Number of pavers across the width
= 200 ÷ 10
= 20
Total pavers = 15 × 20
= 300

**11** Perimeter = 36 cm
All 4 sides are the same length.
Length of each side = 36 cm ÷ 4
= 9 cm
Area = 9 cm × 9 cm
= 81 $cm^2$

**12** Find the volume of each prism.
First prism:
$V = l \times b \times h$
$= 7 \times 5 \times 4$
$= 140$

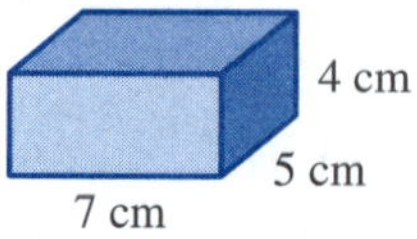


The volume of the first prism is 140 $cm^3$.
Second prism:
$V = l \times b \times h$
$= 8 \times 4 \times 4$
$= 128$

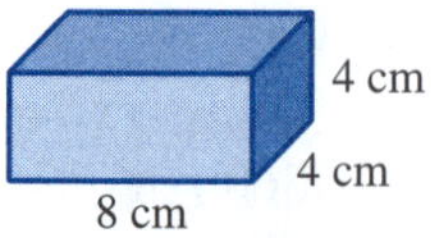


The volume of the second prism is 128 $cm^3$.
Third prism:
$V = l \times b \times h$
$= 3 \times 5 \times 9$
$= 135$

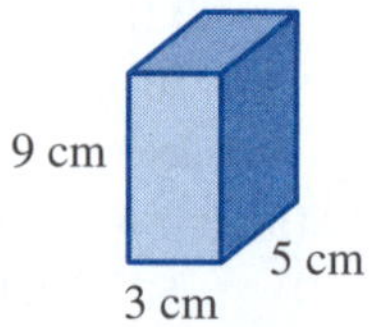


The volume of the third prism is 135 $cm^3$.
Fourth prism:
$V = l \times b \times h$
$= 3 \times 4 \times 10$
$= 120$

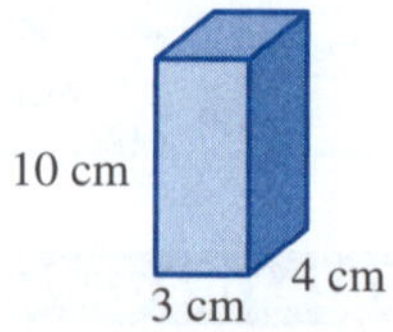


The volume of the fourth prism is 120 $cm^3$.
In order from the least volume to the greatest volume, the prisms are:

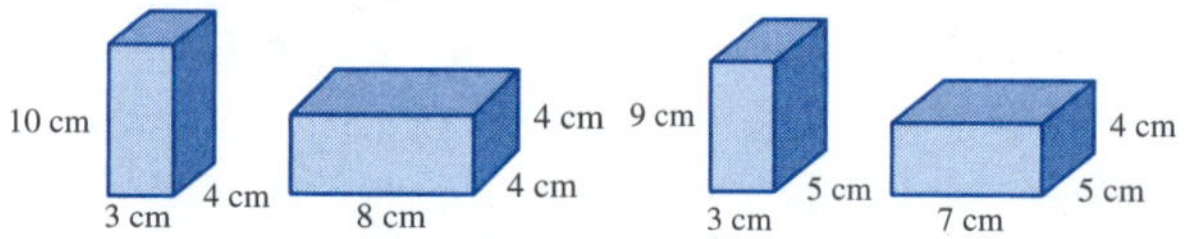


## Intermediate level questions

**Mini Test 18:** Ratio and Rates .................. Page 18

**1** B **2** 60 L **3** C **4** C **5** 600 km **6** D **7** A **8** A
**9** 120 L **10** D **11** B **12** A **13** 39 **14** 54 km
**15** B **16** 800 km **17** $190 **18** C **19** D **20** 6 L

**1** For every 7 blue pens there are 2 red pens.
Number of lots of 7 blue pens = 42 ÷ 7
= 6
Number of red pens = 6 × 2
= 12

**2** The car uses 8 L for every 100 km.
Now 750 km = 7.5 × 100 km
Petrol used = 7.5 × 8 L
= 60 L

**3** On average, each hour Raad ran 16 km.
In 2.5 hours, distance = 2.5 × 16 km
= 40 km

**4** There were 5 females for every 4 males at the meeting.
So 5 out of 9 people were female.
Number of females = $\frac{5}{9}$ of 180
= 100

# Year 7 Numeracy Mini Test Answers

**5** The car travels 100 km on 9 litres.
Now 54 litres = 6 × 9 litres
So the car will travel 6 × 100 km or 600 km.

**6** 16 paces cover 10 m.
Number of 16 paces in 640 paces
= 640 ÷ 16
= 40
So 640 paces cover 40 × 10 m or 400 m.

**7** For every 36 boys there were 24 girls.
So, for every 3 boys, there were 2 girls.
[After dividing by 12.]
The ratio of boys to girls was 3 to 2.

**8** Amount each week = \$20 × 4
= \$80
Number of weeks = \$560 ÷ \$80
= 7

**9** 1 litre flows every 5 seconds.
So, 12 litres will flow every 60 seconds.
This means 12 litres flow every minute.
Amount in 10 minutes = 10 × 12 L
= 120 L

**10** For every 3 blue marbles, there are 5 red marbles.
So, out of every 8 marbles, 3 are blue.
The fraction of blue marbles is $\frac{3}{8}$.

**11** Guido travelled 210 km in 3.5 hours.
Speed = (210 ÷ 3.5) km/h
= 60 km/h

**12** The car uses 7 L /100 km.
Now 420 km = 4.2 × 100 km
Fuel used = 4.2 × 7 L
= 29.4 L
The best estimate for the fuel used is 30 litres.
[Or, to travel 400 km the car would use 4 × 7 litres or 28 litres. To travel 500 km it would use 5 × 7 or 35 litres. So, of the options, the best estimate for 420 km must be 30 litres.]

**13** Number of hours worked = \$1092 ÷ \$28
= 39

**14** Anjuli travels 72 km every hour.
45 minutes is three-quarters of an hour.
Distance travelled = $\frac{3}{4}$ of 72 km
= 54 km

**15** For every 7 cars there were 2 trucks.
So, 2 out of 9 vehicles were trucks.
Number of trucks = $\frac{2}{9}$ × 504
= 112

**16** The car travels 100 km on 8 litres of fuel.
Now, 64 litres = 8 × 8 litres
So, the car will travel 8 × 100 km or 800 km.

**17** From 9:15 am until 2:15 pm is 5 hours.
Sylvia hires the machine for more than 5 hours, but less than 6. She must pay for 6 hours altogether.
Cost = \$40 + 5 × \$30
= \$40 + \$150
= \$190

**18** Jacob travels 12 km in 20 minutes.
So he would travel 3 times as far in 3 times the minutes.
He would travel 36 km in 60 minutes.
So the average speed is 36 km/h.

**19** Sacha travelled at 60 km/h so she travelled 1 km every minute.
She would travel 105 km in 105 minutes.
Now 105 minutes = 1 h + 45 minutes
= $1\frac{3}{4}$ h

**20** Area to be painted = 48 $m^2$
Area for 2 coats = (48 × 2) $m^2$
= 96 $m^2$
Now 1 litre covers 16 $m^2$.
Number of litres = 96 ÷ 16
= 6

## Intermediate level questions

**Mini Test 19:** Mixed Questions ................ Page 19

**1** C **2** B **3** A **4** D **5** 20 **6** C **7** D **8** B **9** D
**10** 6 **11** B **12** 60 **13** A **14** A
**15** Ben, Jim, Tim, Sam, Dan, Joe **16** B

**1** Consider the options:
A triangular prism has 5 faces and 9 edges.

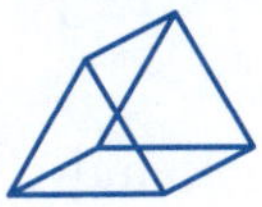

A triangular pyramid has 4 faces and 6 edges.

A rectangular prism has 6 faces and 12 edges.

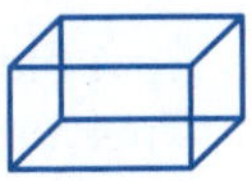

A rectangular pyramid has 5 faces and 8 edges.

The solid that has half as many faces as edges is the rectangular prism (because 6 is half of 12).

**2** $4\frac{2}{3} = \frac{4 \times 3 + 2}{3}$
$= \frac{14}{3}$

**3** $0.35 \div 0.5 = 3.5 \div 5$
$= 0.7$

$5\overline{)3.{}^{3}5}$ = 0.7

**4** 1 kL = 1000 L
1 kL + 30 L = (1000 + 30) L
= 1030 L
= 1.03 kL

**5** 2, 5, 8, 11
The numbers are increasing by 3 each time.
Continue the pattern:
2, 5, 8, 11, 14, 17, 20, 23, 26, …
The seventh number in the pattern is 20.

**6** Total number of ribbons = 5 + 3 + 7
= 15
So 3 out of 15 ribbons are blue.
Chance of blue ribbon $= \frac{3}{15}$
$= \frac{1}{5}$

**7** The first three shapes can be formed from Boyd's two objects.

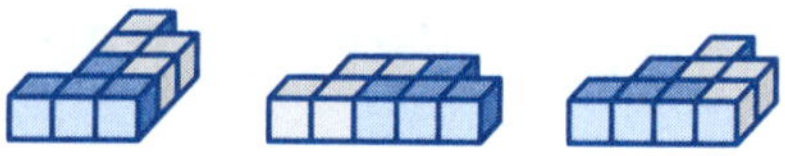

The shape that cannot be formed is 

**8** Total number of roses = 18 + 6
= 24
Fraction of white roses $= \frac{6}{24}$
$= \frac{1}{4}$

**9** Remove one block from each side of the balance.

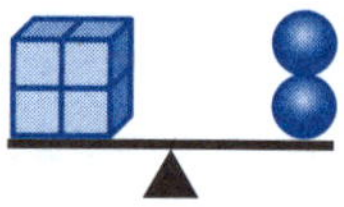

So 4 blocks balance 2 discs.
This means that 2 blocks balance 1 disc.
The number of blocks that balance 3 discs is 3 × 2 or 6.

**10** The answer is 47.
Before adding 5 the answer would have been 47 − 5 or 42.
Now 6 × 7 = 42 so the number that was multiplied by 7 must have been 6.
The number was 6.

**11** An obtuse angle measures more than 90° but less than 180°.

The obtuse angle is 

**12** Shape 1 Shape 2 Shape 3 Shape 4

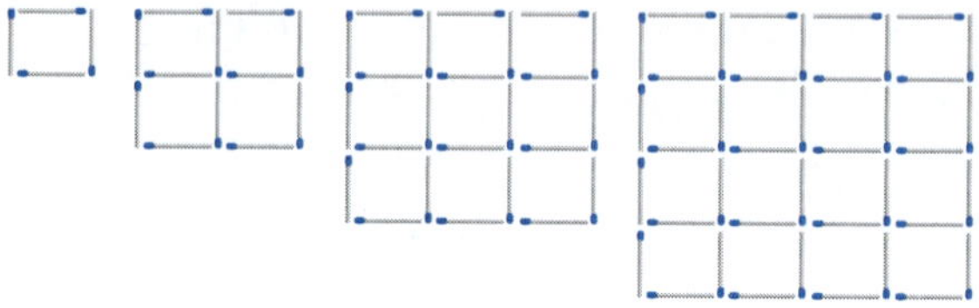

| **Shape** | 1 | 2 | 3 | 4 |
|---|---|---|---|---|
| **Number of matches** | 4 | 12 | 24 | 40 |

The pattern for the number of matches is 4, 12, 24, 40, …
The differences between these numbers are 8, 12 and 16. The differences are increasing by 4 each time. The next difference will be 20.
The number of matches for Shape 5
= 40 + 20
= 60

**13** Write each decimal with 4 digits after the decimal point:
0.2000, 0.0400, 0.0039, 0.0070.
In order, from lowest to highest, the numbers are 0.0039, 0.0070, 0.0400 and 0.2000.
The largest number is 0.2.

**14** The top length is 12 cm so the total length of the bottom sides must be 12 cm.
The total length of the right side is 6 cm so the total length of the left side must also be 6 cm.

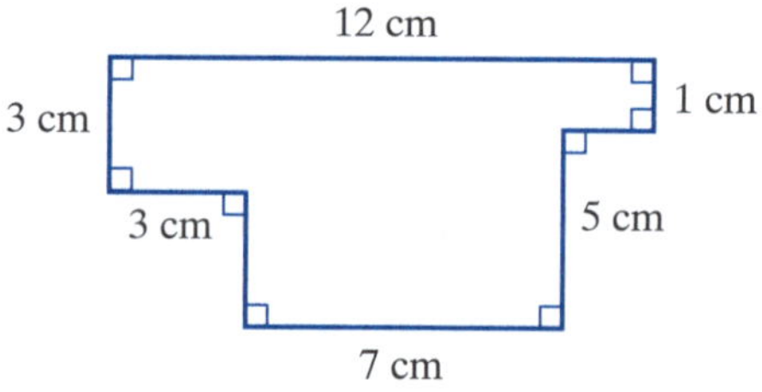


Perimeter = (2 × 12 + 2 × 6) cm
= (24 + 12) cm
= 36 cm

**15** Joe is older than Dan who is older than Sam who is older than Tim (who is older than both Ben and Jim). Jim is older than Ben.
So, from youngest to oldest, the men are Ben, Jim, Tim, Sam, Dan and Joe.

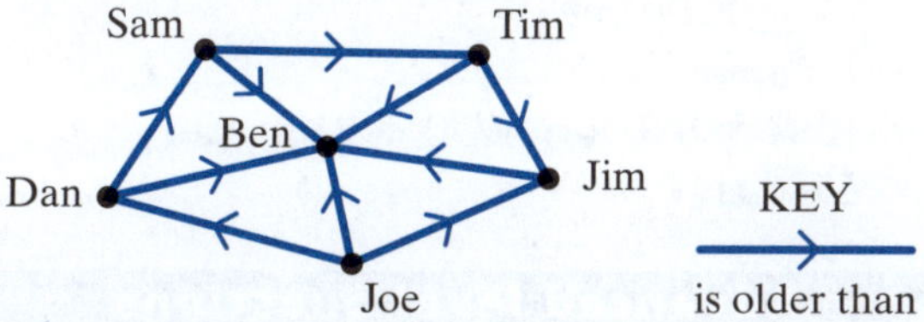


**16** Consider each option:

A: The minute hand on the clock is pointing to 4 and the hour hand is just past 11. The time is twenty minutes past 11 or 11:20. This is not the same as the digital clock, which shows 4:56.

B: The time on the first clock is twenty-five to nine or 8:35. The time on both clocks is 8:35.

C: The time on the first clock is ten to seven or 6:50, which is not the same as the digital clock showing 7:50.

D: The time on the first clock is quarter past two or 2:15 not 2:25.

The time is the same on the two clocks both showing 8:35.

## Intermediate level questions

### Mini Test 20: Mixed Questions ............... Page 20

**1** \$0.65 **2** B **3** 48 **4** 9 L **5** B **6** 128 km **7** C **8** 7 cm **9** −1.5 **10** 84 km **11** 64° **12** C **13** 3.28 **14** D **15** 10 **16** 6776 $cm^2$

**1** Average price per orange = \$15.60 ÷ 24
= \$0.65
The average price per orange was 65 cents.

**2** If Bonnie was 4 years older she would be the same age as Fatima and the sum of the three ages would be 38 + 4 = 42.
If, also, Claire was 3 years younger she would then be the same age as Fatima (and Bonnie). The sum of the three ages would be 42 − 3 = 39.
So, if all 3 were the same age as Fatima the sum of their ages would be 39.
Fatima's age = 39 ÷ 3
= 13

**3** There are 7 divisions on the graph for pink and 2 for orange.
Difference = 5 divisions
So 5 divisions = 20 children
1 division = 4 children
There are 12 divisions for blue.
So number choosing blue = 12 × 4
= 48

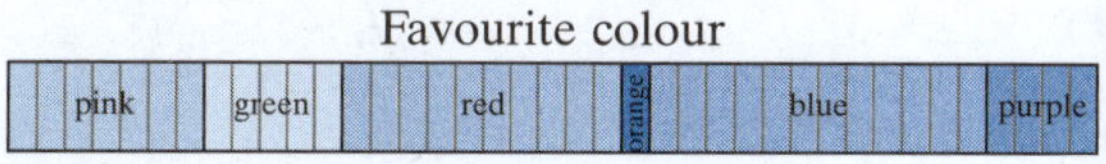


**4** The car uses 7.5 L/100 km.
Litres used in 100 km = 7.5
Litres used in 20 km = 7.5 ÷ 5
= 1.5
Litres used in 120 km = 7.5 + 1.5
= 9
The car will use 9 litres to travel 120 km.

**5** Decrease = 250 − 125
= 125

$$\% \text{ decrease} = \frac{125}{250} \times 100\%$$
$$= \frac{1}{2} \times 100\%$$
$$= 50\%$$

**6** $\frac{1}{8}$ of the distance = 16 km
Distance = 8 × 16 km
= 128 km

**7** There are 3 women for every 5 men. So 3 out of 8 people at the conference are women.

$$\text{Number of women} = \frac{3}{8} \times 120$$
$$= 45$$

**8** Perimeter = 40 cm
So the length plus the width = 20 cm
The length = 13 cm
So the width = (20 − 13) cm
= 7 cm

**9** The arrow is pointing to the number halfway between −2 and −1.

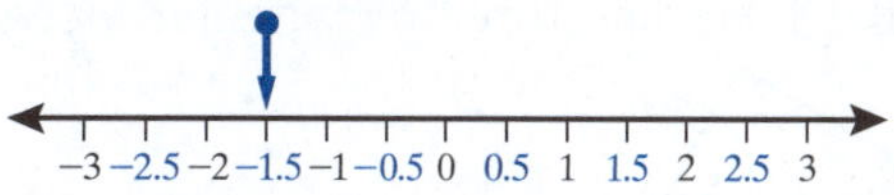


It is pointing to −1.5.

**10** 1 cm represents 15 km.
So 5.6 cm represents 5.6 × 15 km or 84 km.
The actual distance between the two towns is 84 km.

**11** The angle begins on the right-hand side so the scale used on the protractor must be the one that begins with 0° on the right-hand side. This is the inside scale.

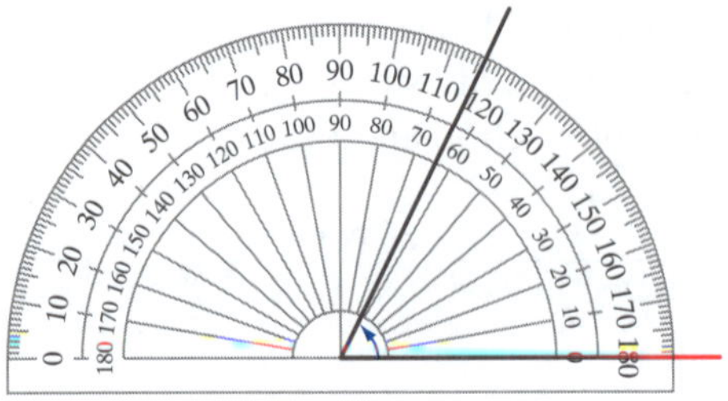

The angle is 64°.

**12** 1 tonne = 1000 kilograms
$8.75 \text{ t} = 8.75 \times 1000 \text{ kg}$
$= 8750 \text{ kg}$

**13** $6.45 + 5.2 = \square + 8.37$
Now $6.45 + 5.2 = 11.65$
So $\square + 8.37 = 11.65$
Now $11.65 - 8.37 = 3.28$
So the number is 3.28.

**14** From 10.54 am until 11.00 am is 6 minutes.
From 11.00 am until 6.00 pm is 7 hours.
From 6.00 pm until 6.09 pm is 9 minutes
Total time = 7 h 15 min

**15** [Look at the pairs of numbers in the table to find the rule connecting them.]

| Number of shapes | 1 | 2 | 3 | 4 |
|---|---|---|---|---|
| Number of pins | 5 | 8 | 11 | 14 |

The number of pins is 3 times the number of shapes plus 2.
So to find the number of shapes for 32 pins first subtract 2 then divide by 3.
Number of shapes $= (32 - 2) \div 3$
$= 30 \div 3$
$= 10$
32 pins will be needed for 10 shapes in the pattern.

**16** The shape is a triangular prism. It has two triangular faces and three rectangular faces.

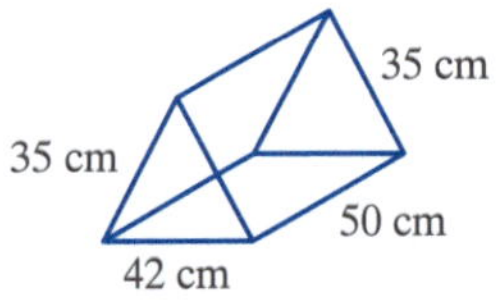


Two rectangles are 35 cm by 50 cm.
Area $= (35 \times 50) \text{ cm}^2$
$= 1750 \text{ cm}^2$
Another rectangle is 42 cm by 50 cm.
Area $= (42 \times 50) \text{ cm}^2$
$= 2100 \text{ cm}^2$
Total area
$= (2 \times 588 + 2 \times 1750 + 2100) \text{ cm}^2$
$= 6776 \text{ cm}^2$

## Advanced level questions

### Mini Test 21: Fractions and Percentages

Page 21

**1** C **2** B **3** C **4** $\frac{2}{3}, \frac{7}{10}, \frac{3}{4}, \frac{4}{5}, \frac{5}{6}$ **5** D **6** B, C
**7** 34% **8** A **9** 18 **10** A **11** D **12** 27 **13** D **14** C
**15** D **16** A

**1** [Write both fractions with a common denominator (15).]
$2\frac{1}{5} = 2\frac{3}{15}$
$2\frac{1}{3} = 2\frac{5}{15}$
Now the number halfway between $2\frac{3}{15}$ and $2\frac{5}{15}$ is $2\frac{4}{15}$.
So the number halfway between $2\frac{1}{5}$ and $2\frac{1}{3}$ is $2\frac{4}{15}$.

**2** $\frac{2}{5}$ are jellies and $\frac{1}{4}$ are mints.
Fraction of jellies and mints $= \frac{2}{5} + \frac{1}{4}$
$= \frac{8}{20} + \frac{5}{20}$
$= \frac{13}{20}$
Remainder $= 1 - \frac{13}{20}$
$= \frac{7}{20}$
The fraction of the sweets that are butterscotch is $\frac{7}{20}$.

**3** $\frac{1}{4}$ of the pets are cats.

Number of cats $= 240 \div 4$
$= 60$

Number of dogs $= 240 - 60$
$= 180$

Number of Labradors $= 30\%$ of 180
$= \frac{3}{10}$ of 180
$= 54$

**4** The set of fractions has denominators 3, 4, 5, 6 and 10.

The common denominator is 60.

$\frac{2}{3} = \frac{40}{60}$

$\frac{3}{4} = \frac{45}{60}$

$\frac{4}{5} = \frac{48}{60}$

$\frac{5}{6} = \frac{50}{60}$

$\frac{7}{10} = \frac{42}{60}$

So, in order, from lowest to highest, the fractions are $\frac{40}{60}, \frac{42}{60}, \frac{45}{60}, \frac{48}{60}, \frac{50}{60}$.

The set of fractions in order is

$\frac{2}{3}, \frac{7}{10}, \frac{3}{4}, \frac{4}{5}, \frac{5}{6}$.

**5** $1\frac{2}{3} \times \frac{4}{5} = \frac{5}{3} \times \frac{4}{5}$

$= \frac{4}{3}$ $\quad \left[\frac{^1\not{5}}{3} \times \frac{4}{\not{5}_1}\right]$

$= 1\frac{1}{3}$

**6** Consider each option:

$\frac{4}{25}$ cannot be simplified.

$\frac{12}{30} = \frac{2}{5}$ [after dividing both numerator and denominator by 6]

$\frac{16}{40} = \frac{2}{5}$ [after dividing both numerator and denominator by 8]

$\frac{28}{80} = \frac{7}{20}$ [after dividing both numerator and denominator by 4]

The fractions that are equal to $\frac{2}{5}$ are $\frac{12}{30}$ and $\frac{16}{40}$.

**7** Total people = 200

| Sport | Cricket | Swimming | Netball | Football |
|---|---|---|---|---|
| Number | 39 | 51 | 42 | |

Number in table $= 39 + 51 + 42$
$= 132$

Number who preferred football $= 200 - 132$
$= 68$

Fraction who preferred football $= \frac{68}{200}$
$= \frac{34}{100}$

So 34% of the people preferred to watch football.

**8** $\frac{13}{4} = 3\frac{1}{4}$

[4 divides into 13 three times with remainder one.]

**9** Number who received top marks is 15% of 180.

Now 10% of 180 = 18

So 5% of 180 = 9

and 15% of 180 = 18 + 9
= 27

Number of girls $= \frac{2}{3}$ of 27
$= 18$

[Or $\frac{2}{3}$ of 15% are girls who received top marks.

Now $\frac{2}{3}$ of 15 is 10, so the number of girls who received top marks is 10% of 180 or 18.]

**10** Increase $= 25\,500 - 25\,000$
$= 500$

% increase $= \frac{500}{25\,000} \times 100\%$
$= \frac{5}{250} \times 100\%$
$= \frac{1}{50} \times 100\%$
$= 2\%$

**11** Fraction $= \frac{50}{320}$
$= \frac{5}{32}$

Now $\frac{1}{3}$ is $\frac{5}{15}$, $\frac{1}{4}$ is $\frac{5}{20}$, $\frac{1}{5}$ is $\frac{5}{25}$ and $\frac{1}{6}$ is $\frac{5}{30}$.

So, of the options, the fraction is closest to $\frac{1}{6}$.

[Or, for each option, find the fraction of 320.

$\frac{1}{3}$ of 320 is more than 100.

$\frac{1}{4}$ of 320 is 80.

$\frac{1}{5}$ of 320 is 64.

$\frac{1}{6}$ of 320 is about 53.]

**12** $\frac{2}{3}$ of a number is 24.

$\frac{1}{3}$ of the number is $24 \div 2$ or 12.

The number is $12 \times 3$ or 36.

Now $\frac{3}{4}$ of 36 is 27.

So $\frac{3}{4}$ of the number is 27.

**13** In 1993, Cameron was 21.

Amy's age in 1993 $= \frac{1}{3}$ of $21 = 7$

Years later $= 2021 - 1993 = 28$

Cameron's age in 2021 $= 21 + 28$
$= 49$

Amy's age in 2021 $= 7 + 28$
$= 35$

Fraction of Cameron's age $= \frac{35}{49}$
$= \frac{5}{7}$

**14** [Change each fraction to an equivalent fraction with a common denominator (24).]

$\frac{1}{3} = \frac{8}{24}$

$\frac{3}{8} = \frac{9}{24}$

$\frac{5}{12} = \frac{10}{24}$

$\frac{7}{24}$

The largest fraction is $\frac{10}{24}$ or $\frac{5}{12}$.

**15** $15 \div \frac{1}{3} = 15 \times 3$
$= 45$

**16** Cups of coconut for 3 dozen $= \frac{3}{4}$

Cups of coconut for 1 dozen $= \frac{3}{4} \div 3$
$= \frac{1}{4}$

Cups of coconut for 4 dozen $= \frac{1}{4} \times 4$
$= 1$

Georgia should use 1 cup of coconut.

## Advanced level questions

### **Mini Test 22:** Decimals, Negatives and Ratio Page 22

**1** D **2** C **3** 7%, 0.077, $\frac{1}{7}$, 0.7 **4** D **5** 5.168 **6** C **7** A **8** 1.6 **9** 13° C **10** A **11** A **12** C **13** B, D **14** D **15** \$3250 **16** B **17** B **18** D **19** 5.1 km **20** −8.8

**1** $77 \div 0.7 = 770 \div 7$
$= 110$

**2** $18 \times 37 = 666$

$1.8 \times 0.37 = 0.666$

[There are three digits, in total, after the decimal points in the question so there must be three digits after the decimal point in the answer.]

**3** [Express each as a decimal.]

7% = 0.07 [$\frac{7}{100}$]

0.7

$\frac{1}{7} = 0.142857\ldots$ [$1 \div 7$]

0.077

In order from lowest to highest, these decimals are 0.07, 0.077, 0.142 857… 0.7

So, in order from lowest to highest, the numbers are 7%, 0.077, $\frac{1}{7}$ and 0.7.

**4** $\frac{13}{100} + \frac{7}{10} = \frac{13}{100} + \frac{70}{100}$
$= \frac{83}{100}$
$= 0.83$

**5** $1.6 + 3.478 + 0.09 = 1.600 + 3.478 + 0.090$
$= 5.168$

$$\begin{array}{r} {}^{1\,1}\phantom{.000} \\ 1.600 \\ 3.478 \\ +\ \underline{0.090} \\ 5.168 \end{array}$$

**6** [The number halfway between two others is the average of the two numbers.]

$0.17 + 0.43 = 0.6$

$$\begin{array}{r} {}^{1}\phantom{00} \\ 0.17 \\ +\ \underline{0.43} \\ 0.60 \end{array}$$

$0.6 \div 2 = 0.3$

The number halfway between 0.17 and 0.43 is 0.3.

**7** $\dfrac{3 - 0.6}{1 + 0.2} = \dfrac{2.4}{1.2}$
$= \dfrac{24}{12}$
$= 2$

**8** 10, 8.6, 7.2, 5.8, …

The numbers are decreasing by 1.4 each time.

The fifth number $= 5.8 - 1.4$
$= 4.4$

The sixth number $= 4.4 - 1.4$
$= 3$

The seventh number $= 3 - 1.4$
$= 1.6$

**9** At midnight the temperature was 21 degrees below zero.

At 8 am the temperature was 8 degrees below zero.

Difference $= 21 - 8$
$= 13$

The temperature had risen 13 degrees.

**10** The ratio of men to women was 5 to 4 so 5 out of every 9 people were men.

Now $72 \div 9 = 8$

So there were 8 lots of 9 people.

Number of men $= 5 \times 8$
$= 40$

Number of women $= 4 \times 8$
$= 32$

Now 5 men and 12 women leave.

New number of men $= 40 - 5$
$= 35$

New number of women $= 32 - 12$
$= 20$

New ratio of men to women = 35 to 20
= 7 to 4

**11** $0.02 \div 0.8 = 0.2 \div 8$
$= 0.025$

$$\begin{array}{r} 0.0\,2\,5 \\ 8\overline{)0.2^{2}0^{4}0} \end{array}$$

**12** Consider each option:

$0.15 + 0.45 = 0.6$

$$\begin{array}{r} {}^{1}\phantom{00} \\ 0.15 \\ +\ \underline{0.45} \\ 0.60 \end{array}$$

$1 - 0.4 = 0.6$

$$\begin{array}{r} {}^{0}\not{1}.{}^{1}0 \\ -\ \underline{0.\ 4} \\ 0.\ 6 \end{array}$$

$0.3 \times 0.2 = 0.06$

[There are a total of 2 digits after the decimal point in the question so there must be 2 digits after the decimal point in the answer.]

$0.12 \div 0.2 = 1.2 \div 2$
$= 0.6$

The expression that does not equal 0.6 is $0.3 \times 0.2$.

**13** Consider each option:

$0.2 < 0.05$ ?

$0.2 = 0.20$

So $0.20 > 0.05$

This option is not correct.

$-0.3 > -0.4$ ?

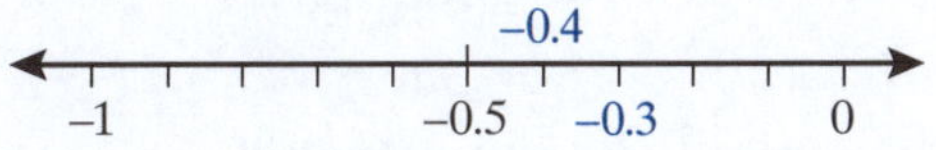


This option is correct.

$-0.6 < -0.7$ ?

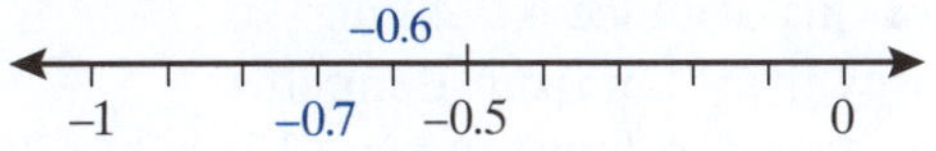


$-0.6 > -0.7$

This option is not correct.

$-0.18 < -0.153$ ?

$-0.180 < -0.153$

This option is correct.

The correct options are $-0.3 > -0.4$ and $-0.18 < -0.153$.

**14**

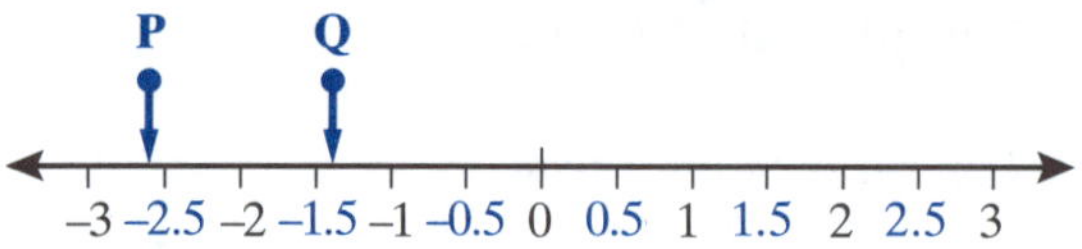


P is between −2.5 and −3.
It is closer to −2.5. P could be −2.6.
Q is between −1 and −1.5.
It is closer to −1.5. Q could be −1.4.
So P could be −2.6 and Q could be −1.4.

**15** Difference = \$2500 − (−\$750)
= \$2500 + \$750
= \$3250

**16** Charlie has 44 chooks.
Before, Charlie had (44 − 12) chooks or 32 chooks.
The ratio of chooks to ducks was 8 to 3.
So for every 8 chooks, Charlie had 3 ducks.
Now 32 ÷ 8 = 4
So Charlie had 4 lots of 8 chooks. He must have also had 4 lots of 3 ducks.
Number of ducks = 4 × 3
= 12
New ratio of chooks to ducks = 44 to 12
= 11 to 3

**17** Each gallon = 4.54 litres
So 2.4 gallons = 2.4 × 4.54 litres
The correct calculation is 2.4 × 4.54.

**18** Number of children = 120
Number of boys = 75
Number of girls = 120 − 75
= 45
Ratio of girls to boys = 45 to 75
= 9 to 15
= 3 to 5

**19** Distance from Bennett to Dunhill
= (7.5 − 2.7) km
= 4.8 km
Half of this distance is 2.4 km.
So Casey is 2.4 km from Bennett.
Distance from Adamstown = (2.7 + 2.4) km
= 5.1 km

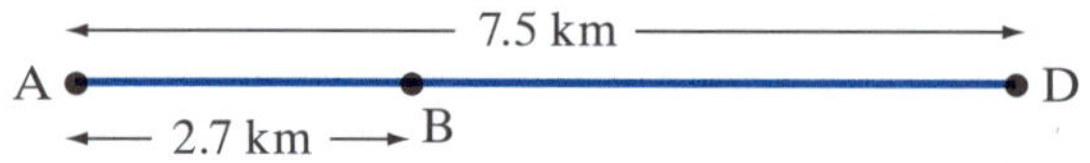


**20** 1.2, −1.3, −3.8, −6.3, …
The numbers in the sequence are decreasing by 2.5 each time.
The next number = −6.3 − 2.5
= −8.8

## Advanced level questions

**Mini Test 23:** Patterns and Algebra....... Page 23

**1** 37 **2** D **3** 37 **4** C **5** C **6** \$80 **7** 32 **8** B **9** 68
**10** A **11** 9801 **12** 8 **13** B **14** 64

**1** 2, 5, 10, 17, 26, ….
The differences between the terms are 3, 5, 7 and 9. The differences are consecutive odd numbers.
The next difference will be 11.
The next number = 26 + 11
= 37

**2**

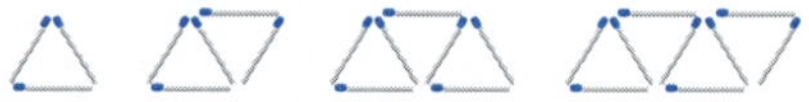

Draw up a table:

| Number of triangles | 1 | 2 | 3 | 4 |
|---|---|---|---|---|
| Number of matches | 3 | 5 | 7 | 9 |

Now consider each option:
'Multiply the number of triangles by 3.'
One triangle:
Number of matches = 3 × 1
= 3 ✓
Two triangles:
Number of matches = 3 × 2
= 6 ✗
The rule is not 'Multiply the number of triangles by 3.'

'Multiply the number of triangles by itself and then add 2.'
One triangle:
Number of matches = 1 × 1 + 2
= 3 ✓
Two triangles:
Number of matches = 2 × 2 + 2
= 6 ✗
The rule is not 'Multiply the number of triangles by itself and then add 2.'

'Multiply the number of triangles by 4 and then subtract 1.'
One triangle:
Number of matches = 4 × 1 − 1
= 3 ✓
Two triangles:
Number of matches = 4 × 2 − 1
= 7 ✗

# Year 7 Numeracy Mini Test Answers

The rule is not 'Multiply the number of triangles by 4 and then subtract 1.'
'Double the number of triangles and then add 1.'
One triangle:
Number of matches $= 2 \times 1 + 1$
$= 3$ ✓
Two triangles:
Number of matches $= 2 \times 2 + 1$
$= 5$ ✓
Three triangles:
Number of matches $= 2 \times 3 + 1$
$= 7$ ✓
Four triangles:
Number of matches $= 2 \times 4 + 1$
$= 9$ ✓
The rule is 'Double the number of triangles and then add 1.'

**3** 1, 2, 4, 7, 11, …
The differences between the terms are 1, then 2, then 3, then 4. The differences are increasing by 1 each time.
Continue the pattern:
$11 + 5 = 16$
$16 + 6 = 22$
$22 + 7 = 29$
$29 + 8 = 37$
The ninth number is 37.

**4** 'Double the number and subtract 3.'
The first number is 4.
The second number $= 4 \times 2 - 3$
$= 5$
The third number $= 5 \times 2 - 3$
$= 7$
The fourth number $= 7 \times 2 - 3$
$= 11$
5, 7 and 11 are all numbers in the pattern.
Of the choices, the number that is not in the pattern is 9.

**5** [Study the pairs of numbers to find the rule connecting them.]

| First number | 1 | 2 | 4 | 7 | 12 |
|---|---|---|---|---|---|
| Second number | 19 | 18 | 16 | 13 | 8 |

The first number plus the second number is always equal to 20.
Now $15 + 5 = 20$
So, when the first number is 15, the second number will be 5.

**6** 'Cost = \$960 divided by the number of people travelling.'
When 12 people travel:
Cost $= \$960 \div 12$
$= \$80$

**7** 'Cost = \$960 divided by the number of people travelling.'
If the cost per person is \$30:
\$30 = \$960 divided by 'number of people'
Now $\$960 \div \$30 = 32$
So $\$960 \div 32 = \$30$
The number of people is 32.

**8** ♦♥♠♣♣♠♥♦♦♥♠♣♣♠♥♦♦♥♠
There are 8 shapes before the pattern begins to repeat.
So every eighth shape will be ♦
Now $6 \times 8 = 48$ so the 48th shape will be ♦
The 49th shape is the start of the new pattern so it is also ♦
The 50th shape will be the second shape in the pattern.
The 50th shape is ♥

**9** 3, 6, 4, 8, 6, 12, 10, 20, 18, …
[The numbers in the pattern first increase then decrease, then increase, then decrease and so on. This suggests that there are two parts to the rule.]
The rule for the pattern is to double the number, then subtract 2, then double that number and then subtract 2 and so on.
The tenth number will be $2 \times 18 = 36$.
The eleventh number will be $36 - 2 = 34$.
The twelfth number will be $2 \times 34 = 68$.

**10** The coordinates of $S$ are $(-1, 0)$.

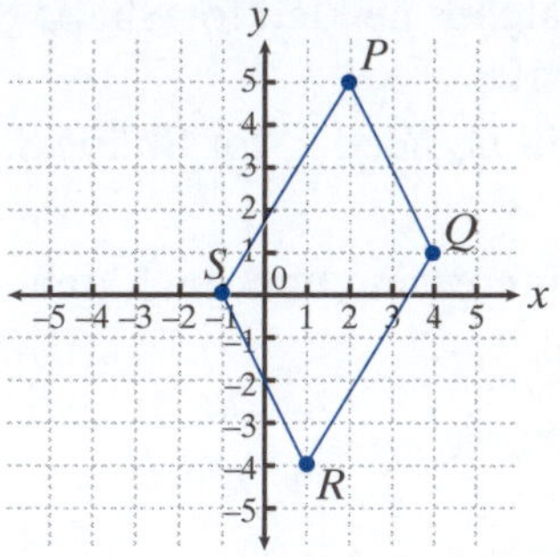


**11** Step 1: $1 \times 1 = 0 \times 2 + 1$
Step 2: $2 \times 2 = 1 \times 3 + 1$
Step 3: $3 \times 3 = 2 \times 4 + 1$
Step 4: $4 \times 4 = 3 \times 5 + 1$
In each step in the pattern the number that is multiplied by itself on the left side is the

same as the step number. The first number on the right side is always 1 less than the step number and the second number on the right side is always 1 more than the step number. The final number on the right side is always 1.

So Step 99 would be:
$99 \times 99 = 98 \times 100 + 1$
So $99 \times 99 = 9800 + 1$
$= 9801$

**12** 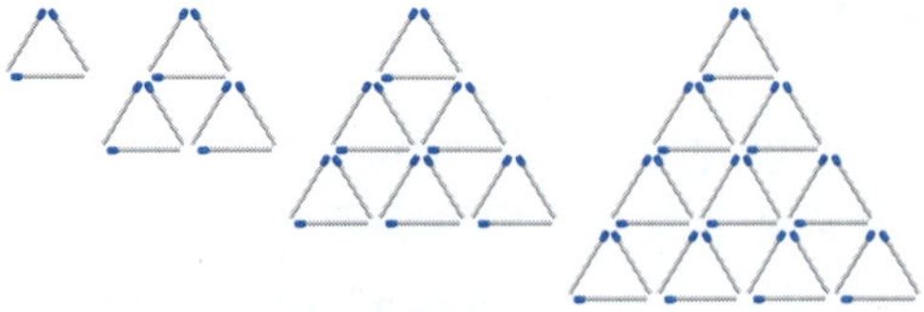

| Shape | 1 | 2 | 3 | 4 |
|---|---|---|---|---|
| Number of matches | 3 | 9 | 18 | 30 |

The pattern for the number of matches is 3, 9, 18, 30, …

The differences between the terms are 6, then 9, then 12. The differences are increasing by 3 each time.

The next difference will be 15.

The number of matches needed for Shape 5 will be $30 + 15 = 45$.

The next difference will then be 18.

The number of matches needed for Shape 6 will be $45 + 18 = 63$.

Continue the pattern.

The next difference will be 21.

The number of matches needed for Shape 7 will be $63 + 21 = 84$.

The next difference will then be 24.

The number of matches needed for Shape 8 will be $84 + 24 = 108$.

So 108 matches will be needed for Shape 8.

**13** 2, 5, 8, 11, 14, …

The numbers are increasing by 3 each time.
The first number is $3 - 1$.
The second number is $2 \times 3 - 1$.
The third number is $3 \times 3 - 1$.
The fourth number is $4 \times 3 - 1$.
The fifth number is $5 \times 3 - 1$.
So the 101st number $= 101 \times 3 - 1$
$= 303 - 1$
$= 302$

**14** [Draw up a table.]

| Shape | 1 | 2 | 3 | 4 |
|---|---|---|---|---|
| Number of pins | 4 | 7 | 10 | 13 |

The number of pins increases by 3 each time.
Number of pins for Shape 1 $= 3 + 1$
Number of pins for Shape 2 $= 2 \times 3 + 1$
Number of pins for Shape 3 $= 3 \times 3 + 1$
Number of pins for Shape 4 $= 4 \times 3 + 1$
So, using this pattern,
Number of pins for Shape 21 $= 21 \times 3 + 1$
$= 63 + 1$
$= 64$

## Advanced level questions

**Mini Test 24:** Space ........................................ Page 24

**1** B **2** A **3** A **4** D **5** A, C **6** C **7** 225° **8** B **9** D **10** 52 cm **11** C **12** D

**1** From the front view you know that the object is 4 blocks across. So the top view cannot be view B because it is at most 3 blocks across. The other choices could be:

The centre object is viewed here from the side.

**2** Consider each choice.

A triangular pyramid has 4 faces and 6 edges.

A triangular prism has 5 faces and 9 edges.

A rectangular pyramid has 5 faces and 8 edges.

A hexagonal prism has 8 faces and 18 edges.

Only the triangular pyramid has its number of edges half as many again as the number of faces. [$\frac{1}{2}$ of 4 is 2; $4 + 2 = 6$]

# Year 7 Numeracy Mini Test Answers

**3** 

In the two views of the cube a total of 5 different faces can be seen. The face that cannot be seen must be the one showing a square. The face opposite the blank face cannot be seen in either view, so that must be the face with the square. So Options B and D cannot be correct.
From the second view of the cube you can see that when the line is below the blank face the dot is on the right side. But in Option C the dot would be on the left side. The only possible option is Option A.

**4** A triangular prism has 2 triangular faces and 3 rectangular (or square) faces. George can build a triangular prism.

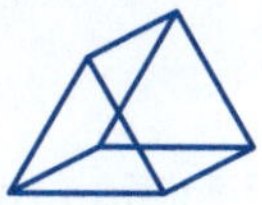

A triangular pyramid has 4 triangular faces. George can build a triangular pyramid.

A square-based pyramid has 1 square face and 4 triangular faces. George can build a square-based prism.

George can build all three objects.

**5** The 6 inside cubes will have no faces painted.

6 cubes in the top layer will have 1 face painted and also 6 in the bottom layer.
3 faces in the front layer and 3 in the back will have 1 face painted.
2 faces on the right side and 2 on the left will have 1 face painted.
Total with 1 face painted = 22
10 cubes in the top layer and 10 in the bottom layer will have 2 faces painted. Another 4 in the middle layer will have 2 faces painted.

Total with 2 faces painted = 24
The 8 corner cubes will have 3 faces painted.
The correct statements are '6 cubes have no blue faces' and '24 cubes have exactly 2 blue faces'.

**6** Numbers on opposite faces add to 7.

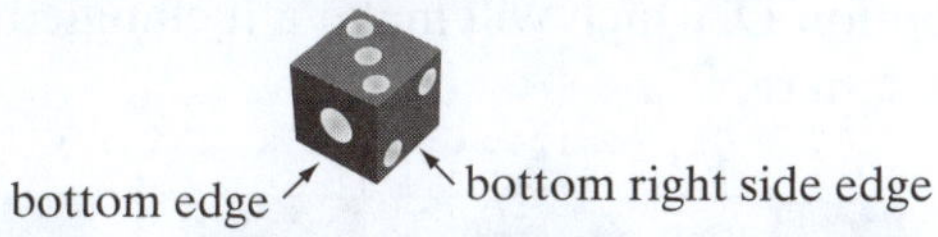


So 6 will be on the back of the original cube. 5 will be on the left side and 4 on the bottom.
When the cube rolls over the bottom edge, 1 will be on the bottom and 6 on the top. The front face will be 3 and the back 4. 2 will still be on the right face and 5 on the left face.

When the cube rolls over the bottom right side edge, 2 will move to the bottom and 5 to the top.
The number on the top face will be 5.

**7** In 2 rotations the arrow turns $1\frac{1}{4}$ times.

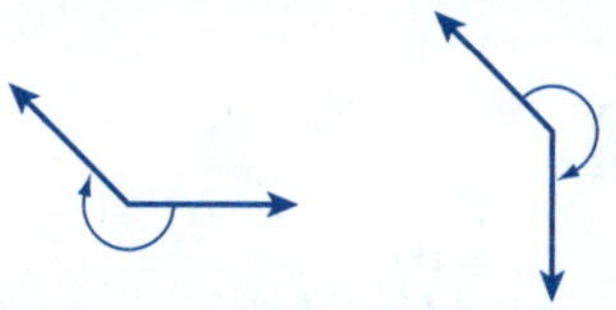

$1\frac{1}{4}$ turns $= 360° + 90°$
$= 450°$

Angle of rotation $= 450° \div 2$
$= 225°$

**8** If a square is cut along a horizontal or vertical axis of symmetry the resulting two parts will be rectangles. The length of the rectangle will be twice the width.

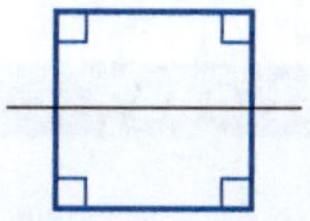

If a square is cut along a diagonal axis of symmetry the resulting two parts will be triangles. These triangles will be both right-angled and isosceles.

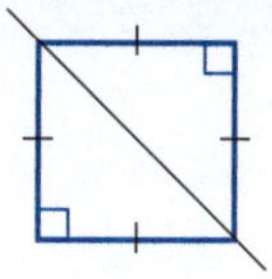

If a square is cut along an axis of symmetry it is not possible to get 2 squares.

**9** Every option will form a square except for option D, which will make a rectangle but not a square.

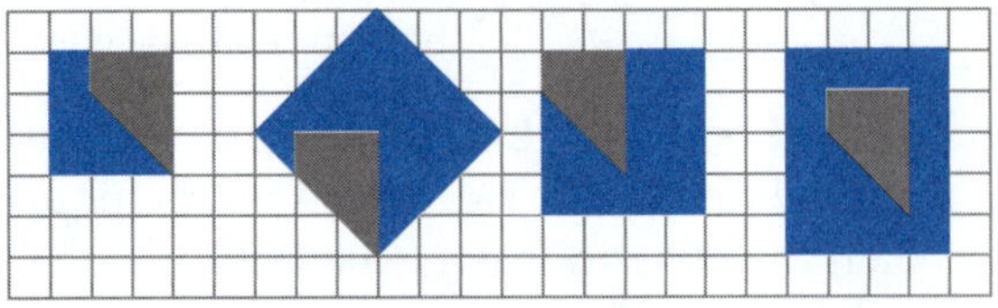

**10** Number of times larger $= 78 \div 12$

$= 6.5$

New width $= 6.5 \times 8$ cm

$= 52$ cm

**11** First the shape is rotated 90° in an anticlockwise direction:

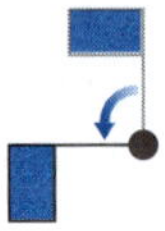

Then it is reflected in a vertical line:

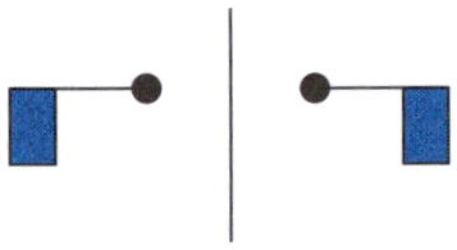

The resulting image is Option C.

**12** The resulting object has 20 edges and 12 faces.

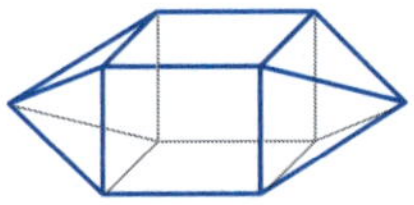

Difference $= 20 - 12$

$= 8$

The object has 8 more edges than faces.

## Advanced level questions

### Mini Test 25: Number Page 25

**1** \$722.50 **2** A **3** 204 **4** A **5** B **6** D
**7** Kelly, Jane, Liz, Sally **8** B **9** 756 **10** C
**11** B **12** 157 **13** 12% **14** D **15** 21 **16** 9.8

**1** Reduction $= 15\%$ of \$850

$= \frac{15}{100} \times \$850$

$= \$127.50$

Price paid $= \$850 - \$127.50$

$= \$722.50$

**2** $\frac{2}{3}$ of $\frac{3}{4}$ of all the cows had male calves.

Fraction $= \frac{2}{3} \times \frac{3}{4}$

$= \frac{1}{2}$

**3** $32\% = 96$

$1\% = 96 \div 32$

$= 3$

$100\% = 100 \times 3$

$= 300$

So there were 300 people at the concert.

Number of adults $= 300 - 96$

$= 204$

**4** Fraction of runs scored by both batsmen combined $= \frac{3}{5} + \frac{1}{3}$

$= \frac{14}{15}$

Fraction that were extras $= 1 - \frac{14}{15}$

$= \frac{1}{15}$

**5** Gain $= \$2500 - \$2000$

$= \$500$

Percentage gain $= \frac{\$500}{\$2000} \times 100\%$

$= \frac{1}{4} \times 100\%$

$= 25\%$

**6** $\frac{3}{5}$ of a number is 45.

$\frac{1}{5}$ of the number is $45 \div 3$ or 15.

The number is $15 \times 5$ or 75.

$\frac{2}{3}$ of $75 = 50$

**7** Liz: Success rate $= \frac{13}{20} \times 100\%$

$= 65\%$

Kelly: Success rate $= \frac{7}{10} \times 100\%$

$= 70\%$

Sally: Success rate $= \frac{9}{15} \times 100\%$

$= 60\%$

Jane: Success rate $= \frac{8}{12} \times 100\%$

$= 66\frac{2}{3}\%$

So, in order from the highest percentage success rate to the lowest, the players are Kelly, Jane, Liz and Sally.

**8** $\dfrac{18.75 + 12.4}{2.4 - 1.6} = \dfrac{31.15}{0.8}$
$= 38.9375$
$= 39$ (nearest whole number)

**9** Number absent = 16% of 900
$= \dfrac{16}{100} \times 900$
$= 144$
Number present = 900 − 144
= 756

**10** [The number halfway between two others is the average.]
$\text{Average} = \dfrac{0.6 + 0.04}{2}$
$= \dfrac{0.64}{2}$
$= 0.32$

**11** The value of the painting is 50% more than it was six years ago so it is 150% of the value six years ago.
150% = $15 000
10% = $1000
100% = $10 000
The value of the painting six years ago was $10 000.

**12** When the number was squared the answer was 24 649.
$\text{Number} = \sqrt{24649}$
$= 157$

**13** Hayden had mowed 60% on Saturday so 40% remained to be mowed.
Percentage mowed on Sunday
= 30% of 40%
$= \dfrac{30}{100} \times 40\%$
= 12%

**14** Consider each option:
$2.75 = 2\dfrac{75}{100}$
$= 2\dfrac{3}{4}$
$2\dfrac{6}{8} = 2\dfrac{3}{4}$ [after dividing both numerator and denominator by 2]
$\dfrac{11}{4} = 2\dfrac{3}{4}$ [4 divides into 11 twice with remainder 3]
$2\dfrac{9}{16}$ cannot be simplified.
The option that is not equal to $2\dfrac{3}{4}$ is $2\dfrac{9}{16}$.

**15** Each day's travel costs $4.50.
Number of days = $98.70 ÷ $4.50
= 21.933…
The most number of days that Lily can travel where the daily fare is cheaper is 21.
[The answer is closer to 22 than 21 but the daily fare would be dearer for 22 days.]

**16** After dividing by 0.7 and multiplying by 3, Frieda's answer was 180.
Before multiplying by 3 her answer would have been 60 (180 ÷ 3).
Before dividing by 0.7 her number would have been 42 (60 × 0.7).
Now Frieda should have divided by 3 first.
42 ÷ 3 = 14
She should then have multiplied by 0.7.
14 × 0.7 = 9.8
Frieda's answer should have been 9.8.

## Advanced level questions

**Mini Test 26:** Patterns and Algebra....... Page 26

**1** 39 063 **2** B **3** A, D **4** 46 811 **5** 23 **6** 19
**7** 458 330 **8** A, B, C **9** 144 **10** 210°C **11** 11
**12** 3016

**1** 1, 3, 13, 63, 313, …
The differences between the terms are 2, 10, 50 and 250. The differences are 5 times bigger each time.
So the rule might involve multiplying the previous number by 5.
Now 1 × 5 = 5
To get 3 you need to subtract 2.
Now 3 × 5 − 2 = 13
and 13 × 5 − 2 = 63
and 63 × 5 − 2 = 313.
The rule is to multiply by 5 and subtract 2.
Continue the pattern:
313 × 5 − 2 = 1563
1563 × 5 − 2 = 7813
7813 × 5 − 2 = 39 063
The eighth number is 39 063.

**2** Try each option:

| △ | 2 | 3 | 5 | 8 |
|---|---|---|---|---|
| ♦ | 7 | 16 | 34 | 61 |

♦ = 6 × △ − 5
When △ = 2,
♦ = 6 × 2 − 5
= 7 ✓
When △ = 3,
♦ = 6 × 3 − 5
= 13 ✗
The rule is not ♦ = 6 × △ − 5.

♦ = 9 × △ − 11
When △ = 2,
♦ = 9 × 2 − 11
= 7 ✓
When △ = 3,
♦ = 9 × 3 − 11
= 16 ✓
When △ = 5,
♦ = 9 × 5 − 11
= 34 ✓
When △ = 8,
♦ = 9 × 8 − 11
= 61 ✓
The rule is ♦ = 9 × △ − 11.

**3** Jo could place her third point at (4, 2) or (1, 4) and make a right-angled triangle.

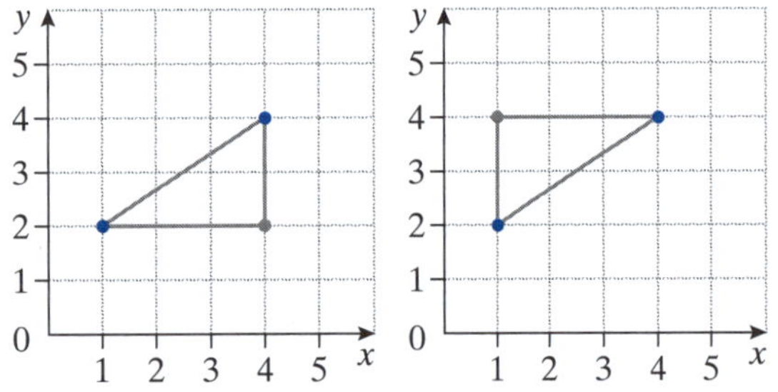


**4** Fourth number = 8 × 731 + 3
= 5848 + 3
= 5851
Fifth number = 8 × 5851 + 3
= 46 808 + 3
= 46 811

**5**

| Number of pentagons | 1 | 2 | 3 |
|---|---|---|---|
| Number of pins | 5 | 8 | 11 |

The number of pins increases by 3 each time.
The rule is to multiply the number of pentagons by 3 and then add 2.
So for 7 pentagons,
Number of pins = 3 × 7 + 2
= 23

**6** Number of pins
= 3 × number of pentagons + 2
Now 60 = 3 × 20
But 2 more pins would be required to make 20 pentagons.
There are only enough pins to make 19 pentagons.

**7** 1, 2, 5, 26, 677, …
The rule is to multiply the number by itself (or square the number) and then add 1.
So 1 × 1 + 1 = 2
2 × 2 + 1 = 5
5 × 5 + 1 = 26
26 × 26 + 1 = 677
Continuing the pattern:
677 × 677 + 1 = 458 329 + 1
= 458 330
The next number is 458 330.

**8**

| Number of squares | 1 | 2 | 3 | 4 |
|---|---|---|---|---|
| Number of matches | | | 8 | |

[The only information you have is that 8 matches are needed for 3 squares.]

Try each option:

'Double the number of squares and add 2.'
2 × 3 + 2 = 8
The rule could be 'Double the number of squares and add 2'.

'Multiply the number of squares by itself and then subtract 1'.
3 × 3 − 1 = 8
The rule could be 'Multiply the number of squares by itself and then subtract 1'.

'Add 5 to the number of squares.'
3 + 5 = 8
The rule could be 'Add 5 to the number of squares'.

'Multiply the number of squares by 4 and subtract 5.'
$4 \times 3 - 5 = 7$ (not 8)
The rule cannot be 'Multiply the number of squares by 4 and subtract 5'.
The rule could be 'Double the number of squares and add 2' or 'Multiply the number of squares by itself and then subtract 1' or 'Add 5 to the number of squares'.

**9** 1, 1, 2, 3, 5, 8, 13
Each number is formed by adding the two previous numbers together.
$(1 + 1 = 2, 1 + 2 = 3, 2 + 3 = 5, 3 + 5 = 8$ and so on.)
Continue the pattern:
$8 + 13 = 21$
$13 + 21 = 34$
$21 + 34 = 55$
$34 + 55 = 89$
$55 + 89 = 144$
The first number greater than 100 is 144.
[The pattern of numbers in this question is known as Fibonacci's sequence.]

**10** The temperature is 410°F.
First subtract 32:
$410 - 32 = 378$
Then divide by 1.8:
$378 \div 1.8 = 210$
The temperature is 210°C.

**11**

| Number of rows | 1 | 2 | 3 | 4 | 5 |
|---|---|---|---|---|---|
| Number of bricks | 1 | 3 | 6 | 10 | 15 |

The number of bricks increases by 2, then 3, then 4, then 5. It is increasing by 1 more each time.
Continue the pattern:
1, 3, 6, 10, 15, 21, 28, 36, 45, 55, 66.
66 is the 11th term of the pattern.
The minimum number of rows needed to stack 66 bricks is 11.

**12** The first number is 8.
The first part of the rule is to square the number.
Now $8 \times 8 = 64$
But the second number is 55.
$64 - 9 = 55$
So the second part of the rule must be to subtract 9.
Third number $= 55 \times 55 - 9$
$= 3025 - 9$
$= 3016$

## Advanced level questions

### Mini Test 27: Perimeter, Area and Volume ........ Page 27

**1** 9 cm **2** C **3** 44 cm **4** A **5** 3 m **6** 136 cm² **7** B **8** 144 cm² **9** 160 **10** 61 m² **11** B **12** 21 600 L **13** 64 m³ **14** 68 cm **15** A, C **16** 28 cm

**1** Area = length × width
So $16 \times \text{width} = 144$
width $= 144 \div 16$
$= 9$
The width is 9 cm.

**2** Each dimension has been halved.
Number of times smaller $= 2 \times 2 \times 2$
$= 8$
Larger volume $= 32\ \text{m}^3$
Smaller volume $= 32\ \text{m}^3 \div 8$
$= 4\ \text{m}^3$

**3** The area is $121\ \text{cm}^2$.
Now $11^2 = 121$
So each side of the square is 11 cm.
Perimeter $= 4 \times 11$ cm
$= 44$ cm

**4** Perimeter = 28 cm
So 2 × (length + width) = 28 cm
length + width = 14 cm
But the width is 5 cm.
So the length = 14 cm − 5 cm
= 9 cm
Area = length × width
= 9 cm × 5 cm
$= 45\ \text{cm}^2$

**5** Volume = length × width × height
So $6 \times 5 \times \text{height} = 90$
$30 \times \text{height} = 90$
height $= 90 \div 30$
$= 3$
The height is 3 m.

**6** Two faces are 8 cm by 3 cm.
Area of each = 8 cm × 3 cm
$= 24\ \text{cm}^2$
Two faces are 8 cm by 4 cm.
Area of each = 8 cm × 4 cm
$= 32\ \text{cm}^2$

Two faces are 3 cm by 4 cm.
Area of each $= 3 \text{ cm} \times 4 \text{ cm}$
$= 12 \text{ cm}^2$
Total area $= 2 \times (24 + 32 + 12) \text{ cm}^2$
$= 2 \times 68 \text{ cm}^2$
$= 136 \text{ cm}^2$

**7** Length $= 2 \times$ width
Perimeter $= 2 \times$ (length + width)
$= 2 \times (2 \times \text{width} + \text{width})$
$= 2 \times (3 \times \text{width})$
$= 6 \times$ width

**8** A regular octagon has 8 equal sides.
So the length of the wire $= 8 \times 6$ cm
$= 48$ cm
A square has 4 equal sides.
Each side $= 48 \text{ cm} \div 4$
$= 12$ cm
Area $= 12 \text{ cm} \times 12 \text{ cm}$
$= 144 \text{ cm}^2$

**9** Each roll is 3 m long and the yard is 12 m wide.
Now $12 \div 3 = 4$
So 4 rolls will fit across the yard in a strip 50 cm wide.
1 m = 100 cm
So 2 strips will cover each metre of the length.
Number of strips $= 20 \times 2$
$= 40$
Total rolls $= 4 \times 40$
$= 160$

**10** Divide the shape into 2 rectangles.

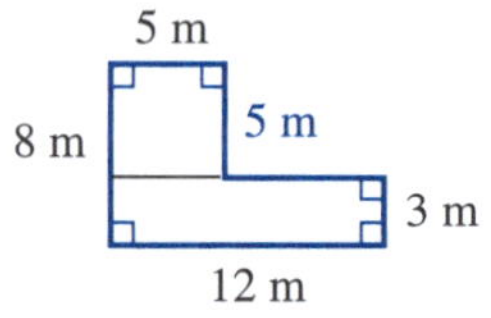


Area $= 5 \text{ m} \times 5 \text{ m} + 12 \text{ m} \times 3 \text{ m}$
$= 25 \text{ m}^2 + 36 \text{ m}^2$
$= 61 \text{ m}^2$

**11** [The units must be the same for both the length and width.]
80 cm = 0.8 m

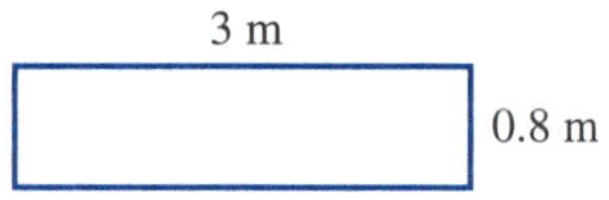


Area = length $\times$ width
$= 3 \text{ m} \times 0.8 \text{ m}$
$= 2.4 \text{ m}^2$

**12** Volume = length $\times$ width $\times$ height
$= 6 \text{ m} \times 3 \text{ m} \times 1.2 \text{ m}$
$= 21.6 \text{ m}^3$
Each cubic metre holds 1000 litres.
Capacity $= 21.6 \times 1000$ L
$= 21\,600$ L

**13** A cube has 6 identical faces.
Area of each face $= 96 \text{ m}^2 \div 6$
$= 16 \text{ m}^2$
Each side $= \sqrt{16}$ m
$= 4$ m
Volume $= 4 \text{ m} \times 4 \text{ m} \times 4 \text{ m}$
$= 64 \text{ m}^3$

**14** Area $= 289 \text{ cm}^2$
Each side $= \sqrt{289}$ cm
$= 17$ cm
Perimeter $= 4 \times 17$ cm
$= 68$ cm

**15** Volume = length $\times$ width $\times$ height
$= 5 \text{ m} \times 4 \text{ m} \times 3 \text{ m}$
$= 60 \text{ m}^3$
Now consider each option:
$6 \text{ m} \times 5 \text{ m} \times 2 \text{ m} = 60 \text{ m}^3$
$9 \text{ m} \times 3 \text{ m} \times 2 \text{ m} = 54 \text{ m}^3$
$15 \text{ m} \times 2 \text{ m} \times 2\text{m} = 60 \text{ m}^3$
$7 \text{ m} \times 3 \text{ m} \times 3 \text{ m} = 63 \text{ m}^3$
The dimensions of the prism could be $6 \text{ m} \times 5 \text{ m} \times 2 \text{ m}$ or $15 \text{ m} \times 2 \text{ m} \times 2 \text{ m}$.

**16** Area = length $\times$ width
So length $\times$ 6 cm $= 48 \text{ cm}^2$
length $= 48 \text{ cm}^2 \div 6$ cm
$= 8$ cm
Perimeter $= 2 \times (8 \text{ cm} + 6 \text{ cm})$
$= 2 \times (14 \text{ cm})$
$= 28$ cm

## Advanced level questions

**Mini Test 28:** Angles ........................................ Page 28

**1** C **2** A **3** D **4** B **5** 120° **6** 80° **7** 110 **8** A **9** C
**10** A **11** D **12** 30 **13** D **14** 135 **15** 60 **16** 84

# Year 7 Numeracy Mini Test Answers

**1** [A polygon can be divided into triangles by drawing lines from one vertex to all the other vertices. The number of triangles is always 2 less than the number of sides. The angles of these triangles add to 180°.]

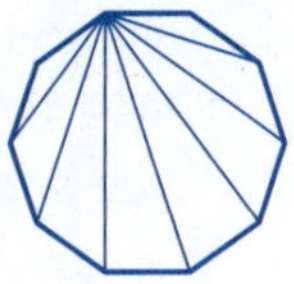

Consider the options:

A hexagon has 6 angles.
Sum of angles $= (6 - 2) \times 180°$
$= 4 \times 180°$
$= 720°$
But $6 \times 144° = 864°$
So the polygon is not a hexagon.

An octagon has 8 angles.
Sum of angles $= (8 - 2) \times 180°$
$= 6 \times 180°$
$= 1080°$
But $8 \times 144° = 1152°$
So the polygon is not an octagon.

A decagon has 10 angles.
Sum of angles $= (10 - 2) \times 180°$
$= 8 \times 180°$
$= 1440°$
Now $10 \times 144° = 1440°$
So the polygon is a decagon.

**2** Angles in a straight line add to 180°.
So $b + c = 180°$
So $a + b + c + d$ must add to more than 180°.
The expression that is not equal to 180° is $a + b + c + d$.

**3** Angles at a point add to 360°.
Consider the options.
A hexagon has 6 angles.
Sum of angles $= (6 - 2) \times 180°$
$= 4 \times 180°$
$= 720°$
Each angle of a regular hexagon $= 720° \div 6$
$= 120°$
Now $3 \times 120° = 360°$
So 3 regular hexagons can meet and fill the space at a point.

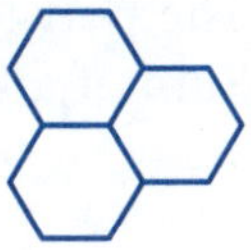

Each angle of a regular hexagon is 120°.
Each angle of an equilateral triangle is 60°.
Each angle of a square is 90°.
Now $120° + 60° + 2 \times 90° = 360°$
So a regular hexagon, an equilateral triangle and 2 squares can meet and fill the space at a point.

An octagon has 8 angles.
Sum of angles $= (8 - 2) \times 180°$
$= 6 \times 180°$
$= 1080°$
Each angle of a regular octagon $= 1080° \div 8$
$= 135°$
Now $2 \times 135° + 90° = 360°$
So 2 regular octagons and a square can meet and fill the space at a point.

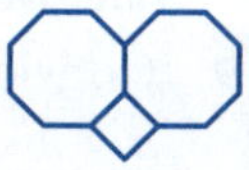

$135° + 2 \times 60° + 90° = 345°$
So a regular octagon, 2 equilateral triangles and a square will not meet and fill the space at a point.

**4** The exterior angle of a triangle is equal to the sum of the interior opposite angles.
So $120° = 70° + x°$
$x = 120 - 70$
$= 50$

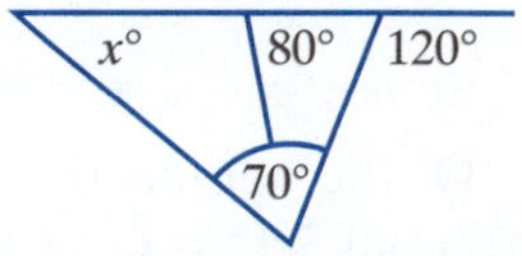


**5** A complete revolution is 360°. So in one hour, the minute hand of a clock turns through 360°. The face of the clock is divided into 12 sections.
So the angle in each section $= 360° \div 12$
$= 30°$
At 8 o'clock there are 4 sections between the two hands.
Angle $= 4 \times 30°$
$= 120°$

**6** The polygon has 5 angles so it is a pentagon.
Sum of angles $= (5 - 2) \times 180°$
$= 3 \times 180°$
$= 540°$
Now 2 angles measure 150°.
Sum of remaining angles $= 540° - 2 \times 150°$
$= 540° - 300°$
$= 240°$
Each angle $= 240° \div 3$
$= 80°$

**7** The angles of a quadrilateral add to 360°.
Sum of given angles $= 75° + 90° + 125°$
$= 290°$
Remaining angle $= 360° - 290°$
$= 70°$
Angles in a straight line add to 180°.
So $x + 70 = 180$
$x = 110$

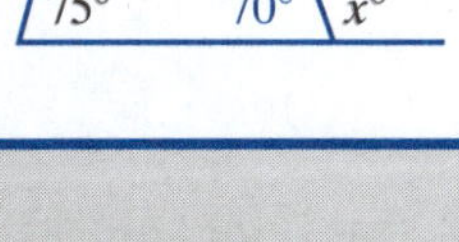

**8** Consider the options.

A pentagon has 5 angles.

$$\begin{aligned}\text{Sum of angles} &= (5 - 2) \times 180^\circ \\ &= 3 \times 180^\circ \\ &= 540^\circ\end{aligned}$$

$$\begin{aligned}\text{Now } 2 \times 90^\circ + 3 \times 120^\circ &= 180^\circ + 360^\circ \\ &= 540^\circ\end{aligned}$$

The polygon is a pentagon.

**9** Consider the options.

$a + b + j = 180^\circ$ (angle sum of a triangle)

So $a + b + j \neq 360^\circ$

$a + b + c + d + e + f + g + h = 360^\circ$ (angle sum of a quadrilateral)

So $d + e + f + g \neq 360^\circ$

$i + j + k + l = 360^\circ$ (angles at a point)

$h + g + i + c + d + k = 360^\circ$ (angles in 2 triangles)

So $h + g + c + d \neq 360^\circ$

Of the options the only expression that must equal $360^\circ$ is $i + j + k + l$.

**10** [Find the remaining angles of the triangle.]

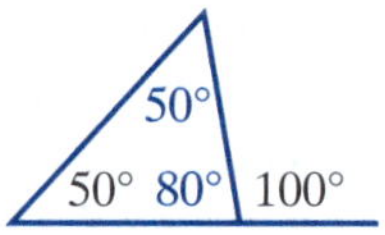


The triangle has 2 equal angles so it is isosceles.

All of the angles are acute.

The triangle is acute-angled and isosceles.

**11** A dodecagon has 12 angles.

$$\begin{aligned}\text{Sum of angles} &= (12 - 2) \times 180^\circ \\ &= 10 \times 180^\circ \\ &= 1800^\circ\end{aligned}$$

Each angle of a regular dodecagon

$= 1800^\circ \div 12$

$= 150^\circ$

Angles at a point add to $360^\circ$.

$$\begin{aligned}\text{Remaining angles} &= 360^\circ - 150^\circ \\ &= 210^\circ\end{aligned}$$

Now consider the options.

A hexagon has 6 angles.

$$\begin{aligned}\text{Sum of angles} &= (6 - 2) \times 180^\circ \\ &= 4 \times 180^\circ \\ &= 720^\circ\end{aligned}$$

$$\begin{aligned}\text{Each angle of a regular hexagon} &= 720^\circ \div 6 \\ &= 120^\circ\end{aligned}$$

Each angle of a square is $90^\circ$.

Now $120^\circ + 90^\circ = 210^\circ$

So a regular hexagon and a square will fill the space at P.

Each angle of an equilateral triangle is $60^\circ$.

$2 \times 60^\circ + 90^\circ = 210^\circ$

So 2 equilateral triangles and a square will fill the space at P.

$150^\circ + 60^\circ = 210^\circ$

So a regular dodecagon and an equilateral triangle will fill the space at P.

$4 \times 60^\circ = 240^\circ$

So 4 equilateral triangles could not be placed beside the dodecagon and fill the space at P.

**12** A hexagon has 6 angles.

$$\begin{aligned}\text{Sum of angles} &= (6 - 2) \times 180^\circ \\ &= 4 \times 180^\circ \\ &= 720^\circ\end{aligned}$$

$$\begin{aligned}\text{Each angle of a regular hexagon} &= 720^\circ \div 6 \\ &= 120^\circ\end{aligned}$$

Each angle of a square $= 90^\circ$.

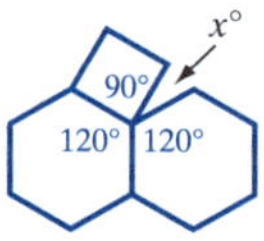


Now angles at a point add to $360^\circ$.

$$\begin{aligned}2 \times 120 + 90 + x &= 360 \\ 330 + x &= 360 \\ x &= 360 - 330 \\ x &= 30\end{aligned}$$

**13** Angles in a straight line add to $180^\circ$ so the third angle of the triangle must be $50^\circ$.

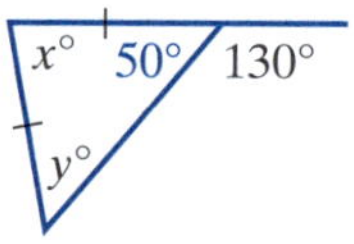


Now the triangle is isosceles. So $y = 50$.

The angles of a triangle add to $180^\circ$.

$$\begin{aligned}\text{So } x + 2 \times 50 &= 180 \\ x + 100 &= 180 \\ x &= 80\end{aligned}$$

$x = 80$ and $y = 50$

**14** The polygon has 5 sides so it is a pentagon.

$$\begin{aligned}\text{Sum of angles} &= (5 - 2) \times 180^\circ \\ &= 3 \times 180^\circ \\ &= 540^\circ\end{aligned}$$

Now 3 angles are right angles.

$$\begin{aligned}\text{Sum of remaining angles} &= 540^\circ - 3 \times 90^\circ \\ &= 540^\circ - 270^\circ \\ &= 270^\circ\end{aligned}$$

So 2 equal angles add to $270^\circ$.

$$\begin{aligned}x &= 270 \div 2 \\ &= 135\end{aligned}$$

**15** Angles in a straight line add to 180°.
So the angle at the top of the triangle is 30°.
But the triangle is isosceles, so another angle is also 30°.

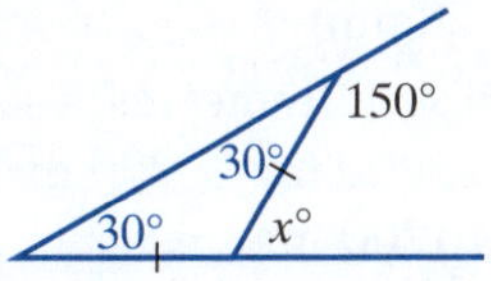


Now, the exterior angle of a triangle is equal to the sum of the interior opposite angles.
So $x = 30 + 30$
$x = 60$

**16** The exterior angle of a triangle is equal to the sum of the interior opposite angles.
The exterior angle $= 76° + 51°$
$= 127°$

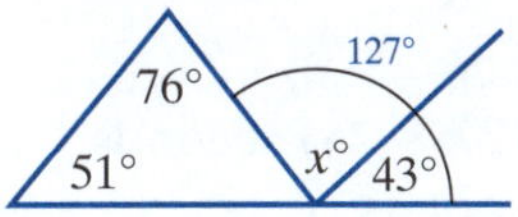


So $x + 43 = 127$
$x = 127 - 43$
$x = 84$

## Advanced level questions

**Mini Test 29:** Rates and Measurements........................................ Page 29

**1** A **2** C **3** 450 mL **4** 750 ml **5** 9.9 km **6** B
**7** 9.4 runs/over **8** C **9** 76 km/h **10** D **11** A
**12** B, E **13** 560 km **14** C

**1** Jeremy left home at 8:10 am.
Time to travel the first 240 km
$= (240 \div 80)$ hours
$= 3$ hours
So Jeremy stopped at 11:10 am.
He stopped for 50 minutes.
So he started again at 12 noon.
From 12 noon until 2:30 pm is $2\frac{1}{2}$ hours.
Distance travelled in afternoon
$= (385 - 240)$ km
$= 145$ km
Speed $= (145 \div 2\frac{1}{2})$ km/h
$= 58$ km/h

**2**

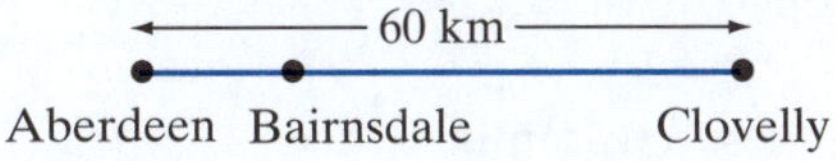


The distance from Aberdeen to Bairnsdale is one part and the distance from Bairnsdale to Clovelly is three parts.
So the distance from Aberdeen to Bairnsdale is $\frac{1}{4}$ of 60 km or 15 km.
Distance from Bairnsdale to Clovelly
$= 3 \times 15$ km
$= 45$ km

**3** On the measuring scale each litre is divided into 5 smaller divisions.
Each division $= 1$ L $\div 5$
$= 1000$ mL $\div 5$
$= 200$ mL
So the bottle held 1200 mL of oil.
Ming removed 750 mL.
Remaining oil $= (1200 - 750)$ mL
$= 450$ mL

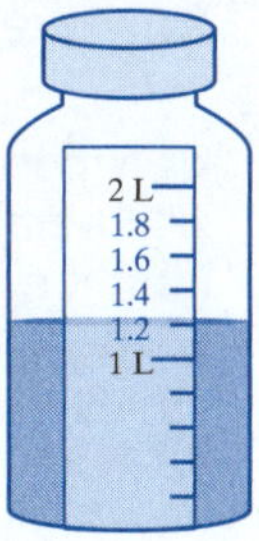


**4** 1 L = 1000 mL
1 kg = 1000 g
Now 1 L weighs 1 kg.
So 1000 mL weighs 1000 g
and 1 mL weighs 1 g.
Weight of water $= (770 - 20)$ g
$= 750$ g
Amount of water $= 750$ mL

**5**

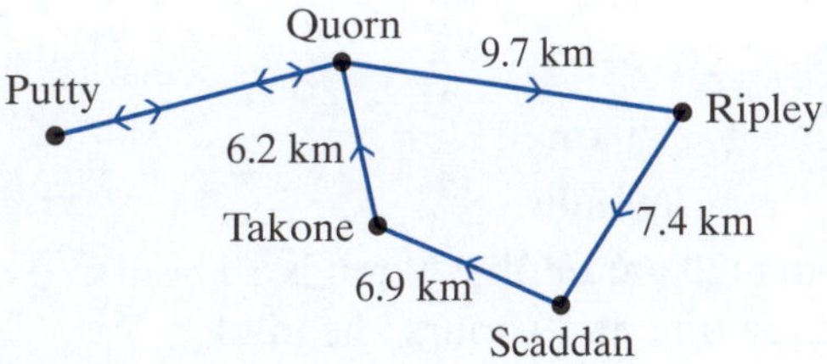


Total distance shown on map
$= (9.7 + 7.4 + 6.9 + 6.2)$ km
$= 30.2$ km
Remaining race distance $= (50 - 30.2)$ km
$= 19.8$ km
This remaining distance is twice the distance from Putty to Quorn.
Distance from Putty to Quorn
$= 19.8$ km $\div 2$
$= 9.9$ km

**6** Time taken $= (594 \div 72)$ h
$= 8.25$ h
$= 8$ h 15 min
The journey began at 7:45 am.
15 minutes after that is 8 am.
8 hours after 8 am is 4 pm.
The journey ended at 4 pm.

**7** 190 runs were required in 20 overs.
After 5 overs, 49 runs were scored.
Remaining runs required $= 190 - 49$
$= 141$
Remaining overs $= 20 - 5$
$= 15$
Run rate = 141 runs in 15 overs
= 9.4 runs/over
[It is not possible to score 9.4 runs in an over. The number of runs must always be a whole number. If a run rate of 9.4 runs per over was needed, then the batting team would not have enough runs if they scored 9 runs every over, but if they scored 10 runs every over they would win before the last available ball had been bowled.]

**8** 1 km = 1000 m
1 m = 100 cm
So 1 km = 100000 cm
So 1 km + 15 cm = (100000 + 15) cm
= 100015 cm

**9** From 8:20 am until 9:00 am is 40 minutes.
It is another 5 minutes until 9:05 am.
From 9:05 am until 12:05 pm is 3 hours.
Time for journey = 3 h 45 min
$= 3\frac{3}{4}$ h
Speed $= 285$ km $\div 3\frac{3}{4}$ h
= 76 km/h

**10** From 5:20 am until 3:20 pm is 10 hours.
Sydney time is 10 hours ahead of London time.
So when it is 10:30 am on Tuesday in Sydney it will be 12:30 am on Tuesday in London.
From 6:45 am until 11:45 am is 5 hours.
London time is 5 hours ahead of New York time.
So when it is 12:30 am on Tuesday in London it will be 7:30 pm on Monday in New York.
The time in New York will be 7:30 pm on Monday.

**11** Time taken $= (42 \div 18)$ h
$= 2.333\ldots$ h
$= 2\frac{1}{3}$ h
$= 2$ h 20 min

**12** 1 hectare = 10000 square metres.
1 ha + 200 $m^2$ = 10000 $m^2$ + 200 $m^2$
= 10200 $m^2$
= 1.02 ha
So both 10200 $m^2$ and 1.02 ha are the same as 1 hectare and 200 square metres.

**13** Number of litres of petrol
= \$63 ÷ \$1.25
= 50.4
Number of kilometres $= 50.4 \div 9 \times 100$
$= 560$
The car will travel 560 km.

**14** Rate = 4 L in 27 seconds
= $(4 \times 25)$ L in $(27 \times 25)$ seconds
= 100 L in 675 seconds
Now 675 s $= (675 \div 60)$ min
$= 11.25$ min
$= 11$ min 15 s

## Advanced level questions

**Mini Test 30:** Chance and Data ............... Page 30

**1** A **2** D **3** 92 **4** C **5** D **6** C **7** D **8** C **9** 100
**10** A **11** US\$60

**1** The average of 7 numbers is 29.
Sum of the numbers $= 29 \times 7$
$= 203$
The average of 6 numbers is 32.
Sum of those numbers $= 32 \times 6$
$= 192$
Difference $= 203 - 192$
$= 11$
The number left out must have been 11.

**2** Total number of cars
$= 16 + 32 + 21 + 12 + 15$
$= 96$
Probability of silver $= \frac{12}{96}$
$= \frac{1}{8}$

| Colour | Number |
|---|---|
| Red | 16 |
| White | 32 |
| Blue | 21 |
| Silver | 12 |
| Other | 15 |

**3** Total of first 6 marks $= 6 \times 78$
$= 468$
To have an average of 80 after 7 tests:
New total $= 7 \times 80$
$= 560$
Difference $= 560 - 468$
$= 92$
Liam would need to score 92 in his seventh test.

**4** The sector graph is divided into 12 equal parts.
5 of those parts represent students with eyes that are brown.

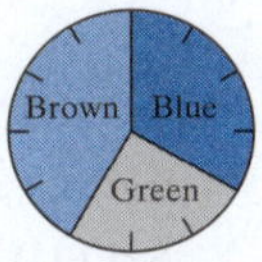


Number of students with eyes that are brown
$= \frac{5}{12} \times 200$
$= 83.333\ldots$
Of the choices the best approximation is 80.

**5** The spinner shows 1 three times, 2 twice and 3 once.
So in 6 spins you would expect to get 1 three times, 2 twice and 3 once.
In 60 spins you would expect to get 1 thirty times, 2 twenty times and 3 ten times.
The table in Option D is the most likely result.

**6** Total of the ages of 11 players
$= 11 \times 32$
$= 352$
Total of the ages of 12 players
$= 352 + 20$
$= 372$
Average age of the 12 players
$= 372 \div 12$
$= 31$

**7** Consider each option.

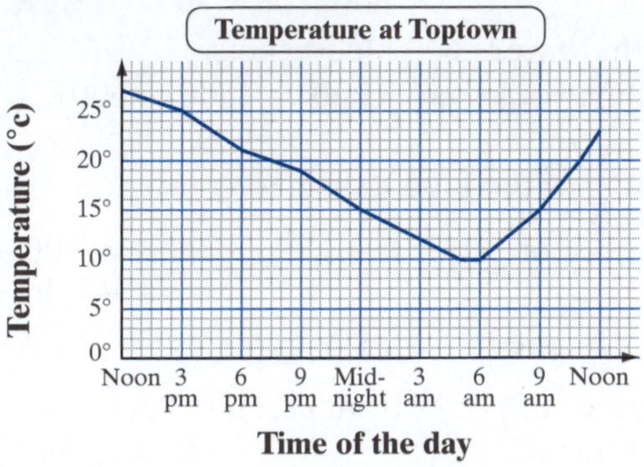


Between noon and midnight:
At noon the temperature was 27°.
At midnight it was 15°.
Difference $= 27° - 15°$
$= 12°$
Between 3 pm and 3 am:
At 3 pm the temperature was 25°.
At 3 am it was 12°.
Difference $= 25° - 12°$
$= 13°$
Between 6 pm and 6 am:
At 6 pm the temperature was 21°.
At 6 am it was 10°.
Difference $= 21° - 10°$
$= 11°$
Between 9 pm and 9 am:
At 9 pm the temperature was 19°.
At 6 am it was 10°
At 9 am it was 15°.
Change $= (19° - 10°) + (15° - 10°)$
$= 9° + 5°$
$= 14°$
The temperature changed the most between 9 pm and 9 am.
[Note: The difference in temperature between 9 pm and 9 am was not the greatest but the change in temperature over that time period was the greatest.]

**8** He was late 30 out of 45 days.
Probability of being late $= \frac{30}{45}$
$= \frac{2}{3}$

**9** The number of men who were passengers
$= 106 - 79$
$= 27$
The total number of passengers is 127.
The number of women who were passengers
$= 127 - 27$
$= 100$

| | Driver | Passenger | Total |
|---|---|---|---|
| Men | 79 | **27** | 106 |
| Women | | **100** | |
| Total | | 127 | |

**10** Sum of ages of male workers
$= 30 \times 25$
$= 750$
Sum of ages of female workers
$= 20 \times 35$
$= 700$
Total of all ages $= 750 + 700$
$= 1450$

Total workers = 30 + 20
= 50
Average age = 1450 ÷ 50
= 29

**11** NZ\$100 = A\$80

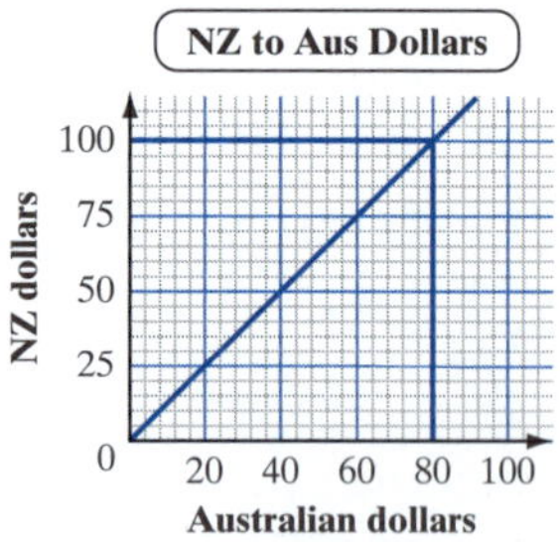


A\$80 = US\$60

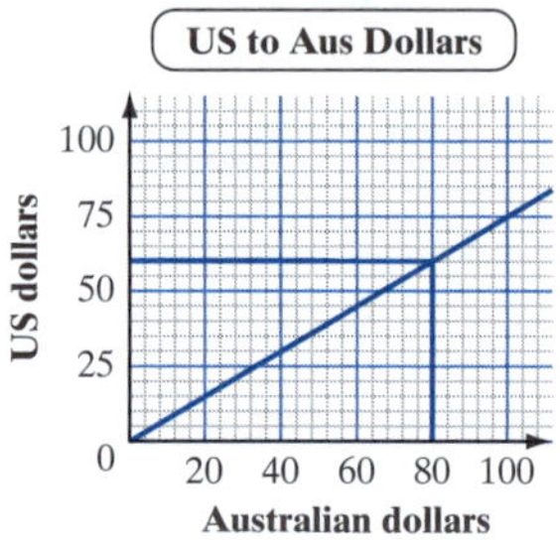


So NZ\$100 = US\$60.

## Advanced level questions

### Mini Test 31: Mixed Questions ............... Page 31

**1** A **2** B **3** 42 cm **4** A, D **5** $\frac{2}{3}, \frac{3}{4}, \frac{7}{9}, \frac{5}{6}$ **6** 425
**7** (0, −6) **8** C **9** 16 **10** 150° **11** D **12** 70 km/h
**13** 61 **14** B **15** −3°C **16** A **17** A, D

**1** $5\frac{3}{4} = \frac{5 \times 4 + 3}{4}$
$= \frac{23}{4}$

**2** The number halfway between two others is the average of those other numbers.
Now $5\frac{1}{2} + 7 = 12\frac{1}{2}$
and $12\frac{1}{2} \div 2 = 6\frac{1}{4}$
So the number halfway between $5\frac{1}{2}$ and 7 is $6\frac{1}{4}$.

**3** Area = length × width
So the length × 9 cm = 108 cm$^2$
Length = (108 ÷ 9) cm
= 12 cm
Perimeter = 2 × (length + width)
= 2 × (12 + 9) cm
= 2 × 21 cm
= 42 cm

**4** $(10 + 3) \times 5 = 13 \times 5$
$= 65$
$(15 - 9) \times (7 + 4) = 6 \times 11$
$= 66$
So $(10 + 3) \times 5 < (15 - 9) \times (7 + 4)$ and $(10 + 3) \times 5 \neq (15 - 9) \times (7 + 4)$
The symbols that make the number sentence correct are $<$ and $\neq$.

**5** The lowest common denominator is 36.
$\frac{2}{3} = \frac{24}{36}$ [after multiplying both numerator and denominator by 12]
$\frac{3}{4} = \frac{27}{36}$ [after multiplying both numerator and denominator by 9]
$\frac{5}{6} = \frac{30}{36}$ [after multiplying both numerator and denominator by 6]
$\frac{7}{9} = \frac{28}{36}$ [after multiplying both numerator and denominator by 4]

So, in order the fractions are $\frac{24}{36}, \frac{27}{36}, \frac{28}{36}, \frac{30}{36}$.
This means that in order from lowest to highest the fractions are $\frac{2}{3}, \frac{3}{4}, \frac{7}{9}, \frac{5}{6}$.

**6** The number of blue tickets is 40% of 1500.
Now 10% of 1500 is 150.
So 40% of 1500 = 4 × 150
= 600
So there were 600 blue tickets.
Number of green tickets = 1500 − 600
= 900
Now $\frac{1}{3}$ of the blue tickets won prizes.
Number of prizes from blue tickets = 600 ÷ 3
= 200
$\frac{1}{4}$ of the green tickets won prizes.
Number of prizes from green tickets
= 900 ÷ 4
= 225
Total prizes = 200 + 225
= 425

# Year 7 Numeracy Mini Test Answers

**7**

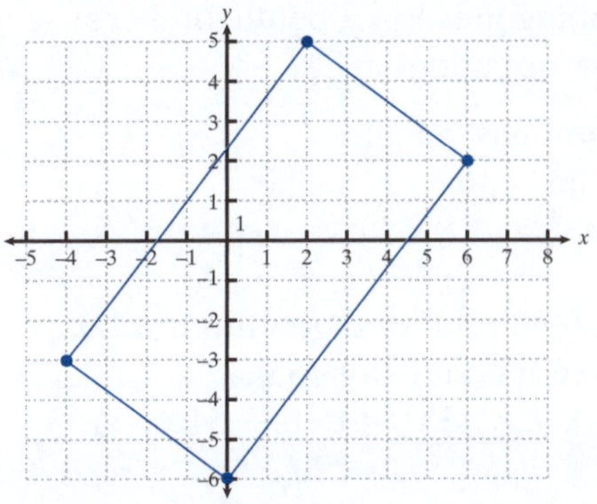

The final point will be at (0, −6).

**8** $0.4 \div 0.8 = 4 \div 8$
$= 0.5$

$$\begin{array}{r} 0.5 \\ 8\overline{)4.{}^{4}0} \end{array}$$

**9** 0.25, 2.5, 4.75, 7, …

$$\begin{array}{r} 2.{}^{4}\not{5}{}^{1}0 \\ -\ \underline{0.25} \\ 2.25 \end{array}$$

$$\begin{array}{r} 4.75 \\ -\ \underline{2.50} \\ 2.25 \end{array}$$

$$\begin{array}{r} {}^{6}\not{7}.{}^{9}\not{1}{}^{1}\not{0}0 \\ -\ \underline{4.75} \\ 2.25 \end{array}$$

The numbers are increasing by 2.25 each time.
Continue the pattern:
0.25, 2.5, 4.75, 7, 9.25, 11.5, 13.75, 16, 18.25, …
The eighth number is 16.

**10** In one hour the minute hand turns through 360°.
There are 12 divisions on the clockface.
Each division = 360° ÷ 12
= 30°
At 5 o'clock there are 5 divisions between the two hands.
Angle between the hands = 5 × 30°
= 150°

**11** Number of red marbles = 120 ÷ 4
= 30
Number of green marbles = 120 ÷ 3
= 40
Total red and green marbles = 30 + 40
= 70
Number of blue marbles = 120 − 70
= 50

Chance of blue marble $= \frac{50}{120}$
$= \frac{5}{12}$

[Or, the chance of red is $\frac{1}{4}$ and the chance of green is $\frac{1}{3}$. The chance of red or green is $\frac{1}{4} + \frac{1}{3} = \frac{3}{12} + \frac{4}{12} = \frac{7}{12}$.
The chance of blue is $1 - \frac{7}{12} = \frac{5}{12}$.]

**12** Distance travelled = 245 km
From 6:20 until 9:20 is 3 hours.
From 9:20 until 9:50 is 30 minutes or $\frac{1}{2}$ hour.
Time taken $= 3\frac{1}{2}$ hours
Speed = 245 km in $3\frac{1}{2}$ h
= 490 km in 7 h
= 70 km/h

**13** First number = 1
Second number = 2 × 1 + 3
= 5
Third number = 2 × 5 + 3
= 13
Fourth number = 2 × 13 + 3
= 26 + 3
= 29
Fifth number = 2 × 29 + 3
= 58 + 3
= 61

**14** For every 5 milk chocolates there are 2 dark chocolates.
So 5 out of every 7 chocolates are milk ones.
Now 35 ÷ 7 = 5
So there are 5 lots of 7 chocolates in the box.
Number of milk chocolates = 5 × 5
= 25
Number of dark chocolates = 5 × 2
= 10
Difference = 25 − 10
= 15
There are 15 more milk chocolates than dark chocolates.

**15** The temperature at midnight = 4 °C
If it was 4° colder the temperature would be zero degrees.

But the temperature dropped by 7° so it dropped another 3°. It was 3 degrees below zero.
The temperature at 5 am was −3 °C.

**16** New prism is half the length, half the width and half the height of the first one.
Number of times smaller $= 2 \times 2 \times 2$
$= 8$
New volume $= (160 \div 8)\ \text{cm}^3$
$= 20\ \text{cm}^3$

**17** $1.2 \times 0.01 = 0.012$
$0.12 \div 0.01 = 12 \div 1 = 12$
$0.012 \times 100 = 1.2$
$12 \div 1000 = 0.012$
So $1.2 \times 0.01$ and $12 \div 1000$ have the same value.

## Advanced level questions

**Mini Test 32:** Mixed Questions ................ Page 32

**1** C **2** 72 **3** 43.9 **4** B **5** 14.4 L **6** A **7** A
**8** B **9** 20 km **10** D **11** 12 **12** A **13** 45 m **14** D

**1** Increase $= 40 - 25$
$= 15$
Percentage increase $= \frac{15}{25} \times 100\%$
$= 60\%$

**2** For every 3 pears there are 8 apples.
Now $192 \div 8 = 24$
So there are 24 lots of 8 apples.
There will also be 24 lots of 3 pears.
Number of pears $= 24 \times 3$
$= 72$

**3** The number halfway between two others is the average of those other numbers.
Now $37.2 + 50.6 = 87.8$
$87.8 \div 2 = 43.9$
So the number halfway between 37.2 and 50.6 is 43.9.

**4** Consider the options.
A pentagon has 5 angles.
Angle sum $= (5 - 2) \times 180°$
$= 3 \times 180°$
$= 540°$
If the polygon is a pentagon there will be 2 angles of 90° and 3 of 135°.
$2 \times 90° + 3 \times 135° = 180° + 405°$
$= 585°$
So the polygon is not a pentagon because the angles do not add to 540°.

A hexagon has 6 angles.
Angle sum $= (6 - 2) \times 180°$
$= 4 \times 180°$
$= 720°$
If the polygon is a hexagon there will be 2 angles of 90° and 4 of 135°.
$2 \times 90° + 4 \times 135° = 180° + 540°$
$= 720°$
So the polygon is a hexagon because the angles do add to 720°.

**5** From 8:40 until 10:40 is 2 hours.
From 10:40 until 11:10 is 30 minutes or half an hour.
So Simon travelled for $2\frac{1}{2}$ hours.
Distance travelled $= 2\frac{1}{2} \times 72$ km
$= 180$ km
Now $180 \div 100 = 1.8$
So Simon travelled for 1.8 lots of 100 km.
Amount of petrol used $= 1.8 \times 8$ L
$= 14.4$ L

**6** 1 kilolitre = 1000 litres
3.9 gigalitres = 3 900 000 000 litres
= 3 900 000 × 1000 litres
= 3 900 000 kilolitres

**7** [Study the pairs of numbers in the table to find the rule connecting them.]

| First number | 1 | 2 | 3 | 4 | 5 |
|---|---|---|---|---|---|
| Second number | | 8 | 27 | 64 | 125 |

The rule is that the second number is the first number cubed.
[$8 = 2 \times 2 \times 2$; $27 = 3 \times 3 \times 3$ etc]
So when the first number is 1,
the second number $= 1 \times 1 \times 1$
$= 1$

**8** Angles in a straight line add to 180°.
So the angle that forms a straight line with the angle of 110° is 70°.
The angle that forms a straight line with the angle of 140° is 40°.
Angles of a triangle add to 180°.
Now 70° + 40° = 110°
Remaining angle = 180° − 110°
= 70°

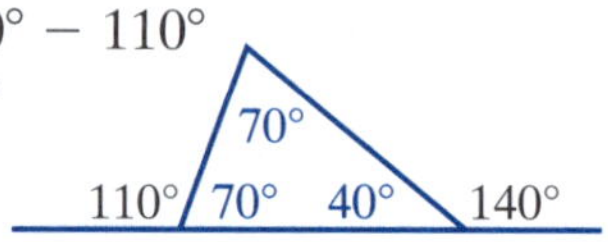

All the angles of the triangle are acute, so it is an acute-angled triangle.
Two of the angles are equal so the triangle is isosceles.
The triangle is acute-angled and isosceles.

**9** Distance = 5 × square root of (height ÷ 2)
= 5 × square root of (32 ÷ 2)
= 5 × square root of 16
= 5 × 4
= 20
The distance to the horizon would be 20 km.

**10** $\frac{\sqrt{15625}}{5 + 20} = \frac{125}{25}$
$= 5$

**11** The average age of 5 children is 8.
Sum of their ages = 8 × 5
= 40
Total of ages of the children and their father
= 40 + 32
= 72
Average age = 72 ÷ 6
= 12

**12** From the side view of the object you can see that it is at most 3 cubes high. The front view showing the object to be 4 cubes high cannot be correct.
The front view cannot be A.
The other views could be given by:

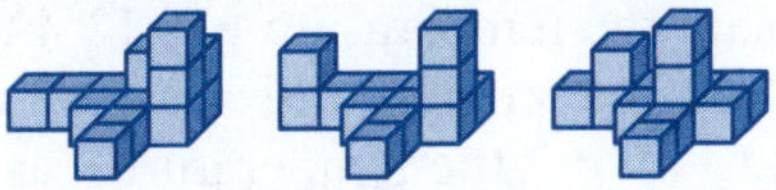

**13** Area = length × width
First rectangle:

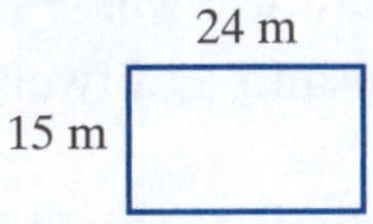


Area = 24 m × 15 m
= 360 $m^2$
So the area of the second rectangle is 360 $m^2$.
Second rectangle:

360 $m^2$ = length × 8 m
length = (360 ÷ 8) m
= 45 m

[Or, the width of the second rectangle is one-third of the length of the first rectangle so, because the areas of the rectangles are the same, the length of the second rectangle must be three times the width of the first rectangle.
3 × 15 = 45]

**14** 35 British pounds is about $80 Australian, which is about $100 New Zealand.

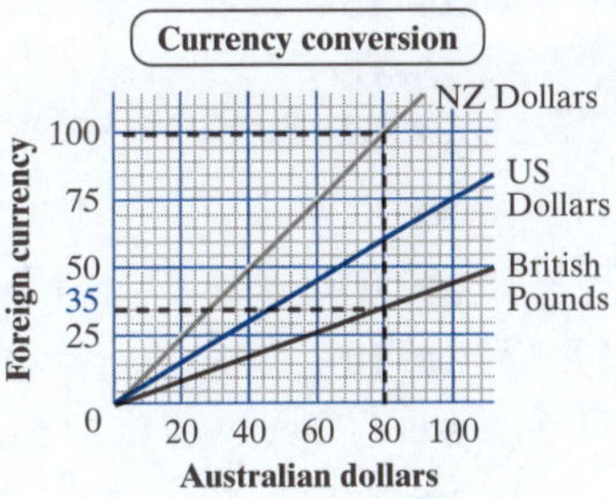

## Standard level

### YEAR 7 Numeracy Test 1

Pages 34–47

**1** C
**2** E
**3** 14
**4** A
**5** A
**6** B
**7** A
**8** −7°C
**9** C
**10** C
**11** B
**12** A
**13** D
**14** B
**15** D
**16** B
**17** C
**18** $0.55
**19** E
**20** −17, −8, −2, 0, 3, 12
**21** A
**22** C
**23** B
**24** A, D
**25** 152°
**26** D
**27** B
**28** C
**29** B
**30** 10
**31** C
**32** A
**33** A
**34** A, D
**35** D
**36** D
**37** 73
**38** C
**39** D
**40** D
**41** E
**42** 28
**43** A, C, D
**44** B
**45** 24
**46** B
**47** B, E
**48** 6, 3, 5, 2

**1** The lamp (at $38.64) is closer to $40 than $30. The printer (at $62.27) is closer to $60 than $70. The chair (at $83.98) is closer to $80 than $90. The best way to estimate the cost is $40 + $60 + $80.

**2** $9 \times 4 = 36$
Try each of the options:
$4 \times 9 = 36$
$18 \times 2 = 36$
$6 \times 6 = 36$
$3 \times 12 = 36$
$5 \times 8 = 40$
The expression that is not equal to $9 \times 4$ is $5 \times 8$.

**3** $6 \times \square = 12 \times 7$
The first number (6) has been doubled so the second number ($\square$) must have been halved to give the same product.
So $\square = 2 \times 7$
$= 14$

**4** $32 \times 85 = 2720$
So $3.2 \times 0.85 = 2.720$
[There are 3 places after the decimal point in the question so there must be 3 places in the answer.]
$3.2 \times 0.85 = 2.72$

**5** $6 \times (2 + 3) - (7 - 4)$
$= 6 \times 5 - 3$
$= 30 - 3$
$= 27$

**6** $3\frac{2}{5} = \frac{3 \times 5 + 2}{5}$
$= \frac{17}{5}$

**7** $2.4 \div 0.2 = 24 \div 2$
$= 12$

**8** At 6 am the temperature was 12° below zero.
At 8 am the temperature was 5° warmer.
As $12 - 5 = 7$, the temperature was 7° below zero at 8 am.
The temperature was −7°C

**9** Eight thousand and twelve $= 8000 + 12$
$= 8012$

**10** The only shape that will look exactly the same when folded along the dotted line is A in option C.

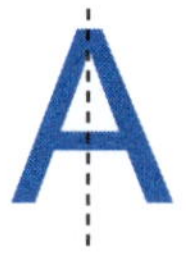

**11** $\frac{18 \times 12}{24 \div 3}$
$= \frac{216}{8}$
$= 27$

**12** The towns North of Kambara are Oxley and Arundel. The towns West of Seaview are Arundel and Newtown.

# Year 7 Numeracy Sample Online-style Test Answers

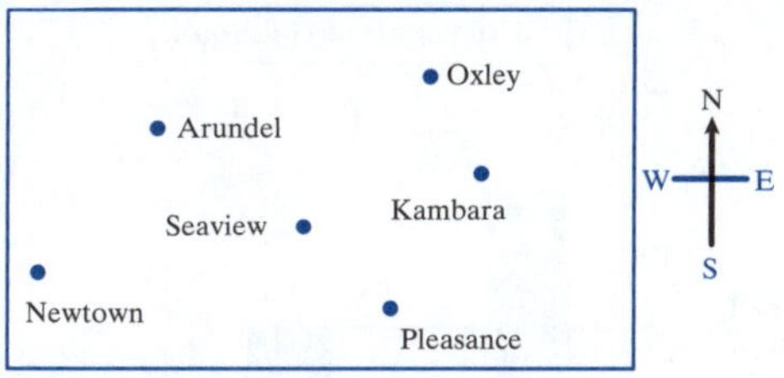


The town that is both North of Kambara and West of Seaview is Arundel.

**13** The shape turns through 90 degrees in a clockwise direction.

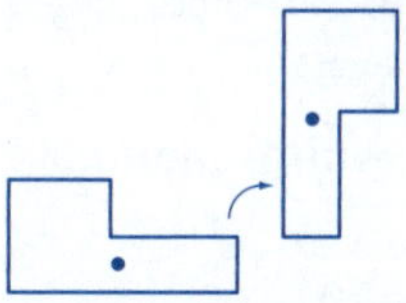

The correct option is D.

**14** 24 m in 3 seconds = (24 ÷ 3) m/s
= 8 m/s

**15** The triangle is divided into 16 smaller triangles. 6 of those smaller triangles are shaded.

Fraction shaded $= \frac{6}{16}$
$= \frac{3}{8}$

**16** Only one box out of six holds more than \$1000. So the chance of choosing that box is one in six. It is unlikely to occur.

**17** Rudi parks in E3.

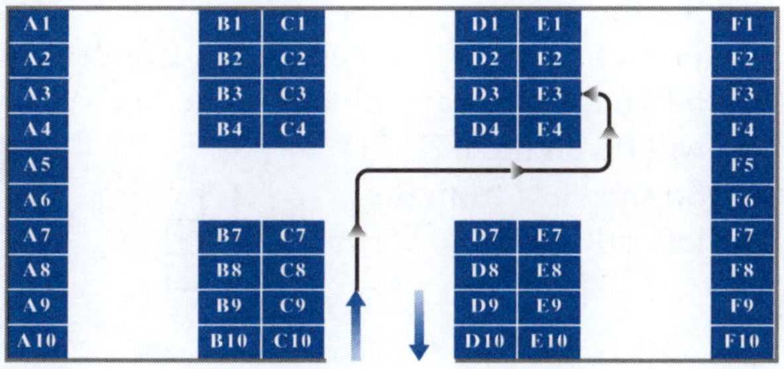


**18** Average price = \$6.60 ÷ 12
= \$0.55

**19** The shape has four sides so it is a quadrilateral. It has two pairs of adjacent sides that are the same length so it is a kite.

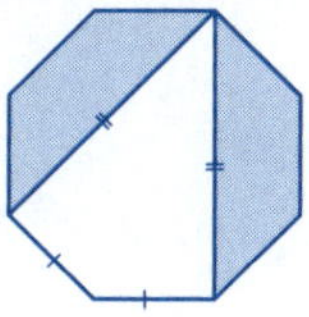

**20** −8, −17 and −2 are less than zero. Of those 3 numbers −17 is the smallest and −2 the largest.

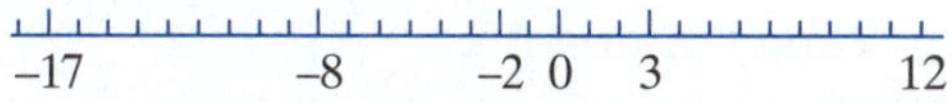


So, in order, the numbers are −17, −8, −2, 0, 3 and 12.

**21** (18.6 + 2.4) ÷ (1.2 − 0.5)
= 21 ÷ 0.7
= 30

**22** An angle of 90° is a right angle, so an angle of 45° is half the size of a right angle. The correct option is C.

**23** The original shape is at most 4 units high. If the height is halved it will be (at most) 2 units high. The original shape is at most 4 units wide. If the width is doubled, it will be (at most) 8 units wide.
The correct option is B.

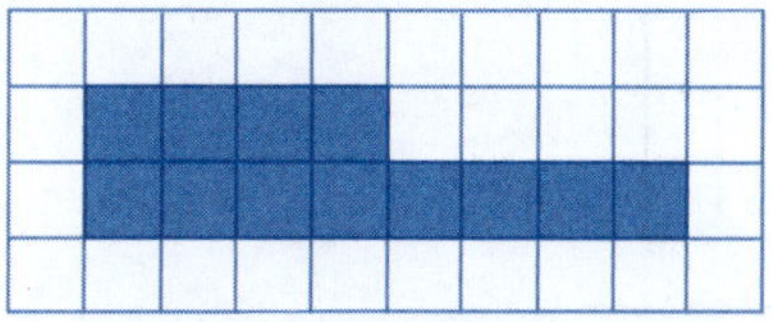

**24** A triangular prism has two triangular faces and three rectangular faces. The correct options are A and D.

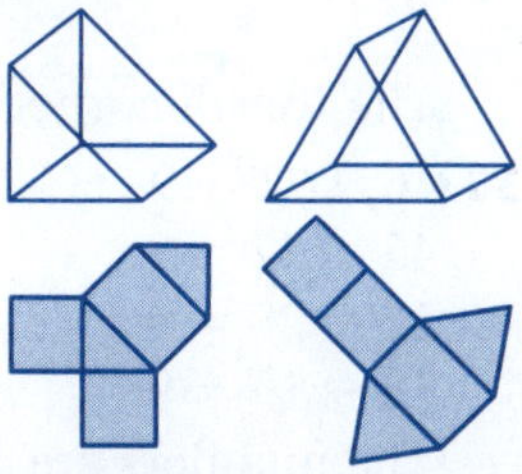

[The net is not option B because the two triangular faces are next to each other, not opposite one another. The net is not C because the faces in option C are all in a line and could not fold to form a solid. It is not option E because that has no rectangular faces. (It is the net of a triangular pyramid.)]

**25** The angle is 152°.

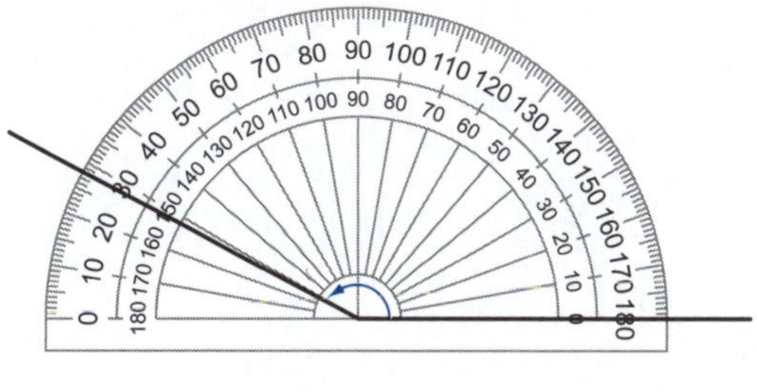

The scale on the protractor must be the inside one, because it begins at 0° where the angle begins.

**26** The correct option is D.

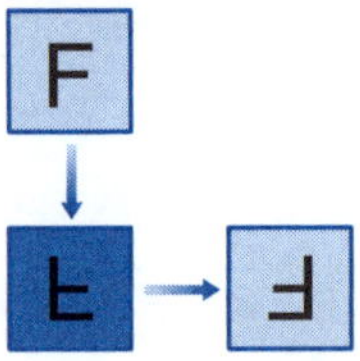


**27** Simone had to wait 20 minutes until 10 o'clock and another 25 minutes after 10 o'clock.

Total time = (20 + 25) min
= 45 min

**28** Saving = 15% of \$180
= \$27

Price paid = \$180 − \$27
= \$153

**29** A third corner is at (1, 1).

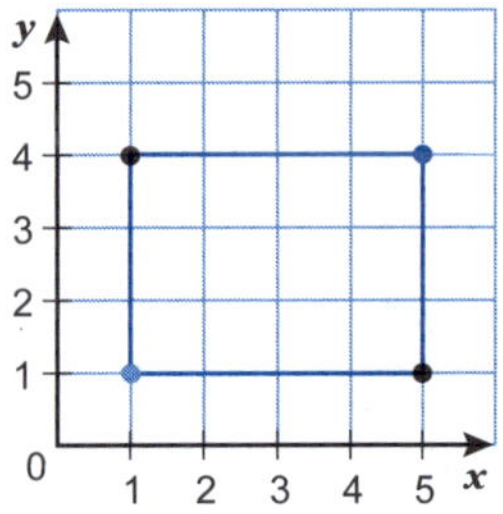


The fourth corner must be at (5, 4).

**30** 46, 37, 28, 19.

46 − 37 = 9
37 − 28 = 9
28 − 19 = 9

The numbers are decreasing by 9 each time.

Next number = 19 − 9
= 10

**31** The graph should show 9 children with brown eyes but shows 8 children.

| Eye colour | Blue | Green | Brown | Other |
|---|---|---|---|---|
| **Number** | 12 | 5 | 9 | 2 |

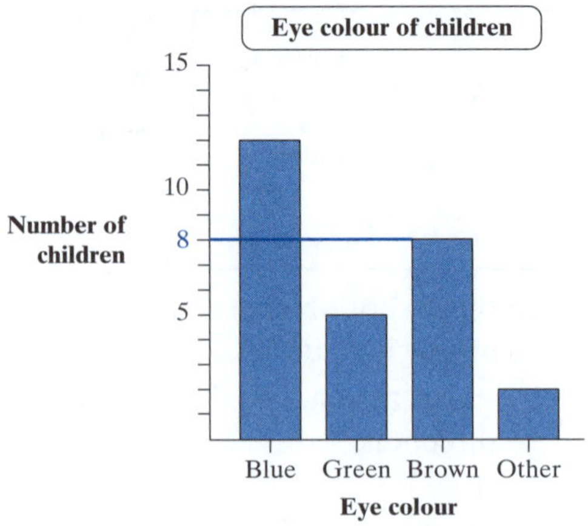


The eye colour with which Anjuli made a mistake is Brown.

**32** Number of females = 32 − 20
= 12

Ratio of females to males = 12 to 20
= 3 to 5

**33** From the front the object is 5 blocks wide and 3 blocks deep.

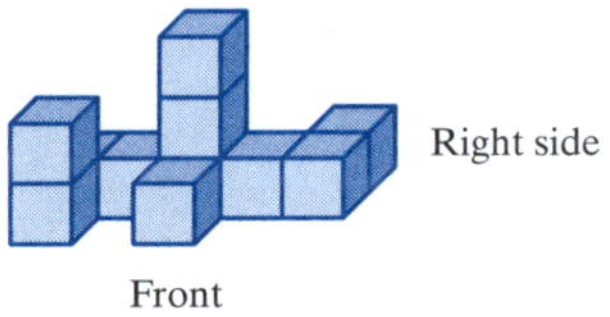


So from the left side the object will be 3 blocks wide.

From the front the blocks are 2 high in the first row, 3 high in the second and 1 high in the third.

From the left side, blocks in the front row will be on the right and blocks in the back row will be on the left.

So the view from the left side is A.

**34** 1 t = 1000 kg
1 kg = 1000 g
1 g = 1000 mg

So the measurements that are the same are 60 kg and 60000 g

[6000000 mg = 6000 g = 6 kg = 0.006 t and 0.6 t = 600 kg = 600000 g = 600000000 mg]

**35** 1 m = 1000 mm

1 m + 30 mm = (1000 + 30) mm
= 1030 mm

# Year 7 Numeracy Sample Online-style Test Answers

**36** Each symbol represents 4 bicycles sold. One-quarter of a symbol will represent 1 bicycle, half a symbol will represent 2 bicycles and three-quarters of a symbol will represent 3 bicycles.

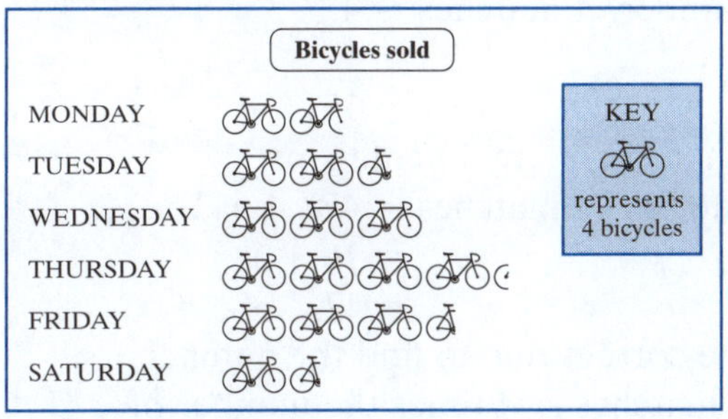


Consider each option.

'There were 19 bicycles sold on Thursday.'

There are 4 and a quarter symbols on Thursday. This represents $4 \times 4 + 1$ or 17 bicycles. This statement is not correct.

'There were 2 more bicycles sold on Friday than on Saturday.'

There are 2 more symbols on Friday than on Saturday but this represents 8 bicycles. This statement is not correct.

'There were 2 fewer bicycles sold on Wednesday than on Tuesday.'

There were 2 more bicycles sold on Wednesday not 2 fewer.

This statement is not correct.

'There were twice as many bicycles sold on Friday than on Monday.'

On Friday there are $3\frac{1}{2}$ symbols.

Number of bicycles sold on Friday
$= 3 \times 4 + 2$
$= 14$

On Monday there are $1\frac{3}{4}$ symbols.

Number of bicycles sold on Monday $= 4 + 3$
$= 7$

Now $14 = 2 \times 7$

So this statement is correct.

The correct statement is 'There were twice as many bicycles sold on Friday than on Monday.'

**37** Consider the options:
The analogue clock in A shows 5 o'clock while the digital clock shows 12:25.
The analogue clock in B shows a quarter past 7 or 7:15 while the digital clock shows 8:15.
The analogue clock in C shows ten minutes to 2 o'clock or 1:50 and the digital clock also shows 1:50.
The analogue clock in D shows 25 minutes to 9 o'clock or 8:35 while the digital clock shows 9:25.
The two clocks in C show the same time.

**38** □ = 3 and ● = 4
□ + ○ = ● + ●
Now ● + ● = 4 + 4
= 8

So □ + ○ = 8

This means 3 + ○ = 8.

So ○ = 5 because 3 + 5 = 8.

**39** Weight of Mitchell's letter $= 60 \times 5$ g
$= 300$ g
So the weight is between 250 g and 500 g.
The cost to post the letter is \$5.00.

**40** The number multiplied by itself (or squared) is 5329.
Number $= \sqrt{5329}$
$= 73$

**41** Angles on a straight line add to 180°.
So $x + 105 = 180$
$x = 180 - 105$
$= 75$

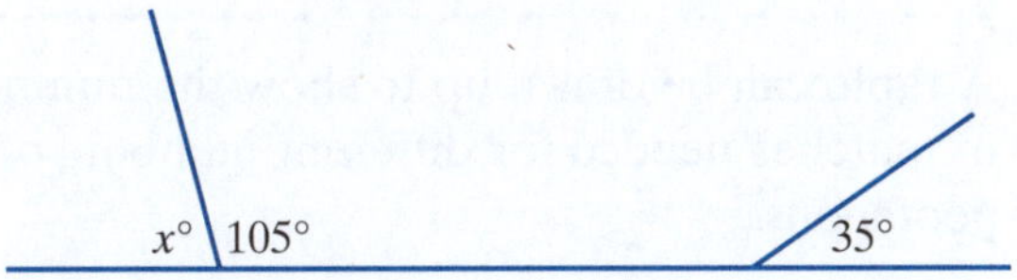


[The angle of 35° has no effect on the size of x.]

**42**

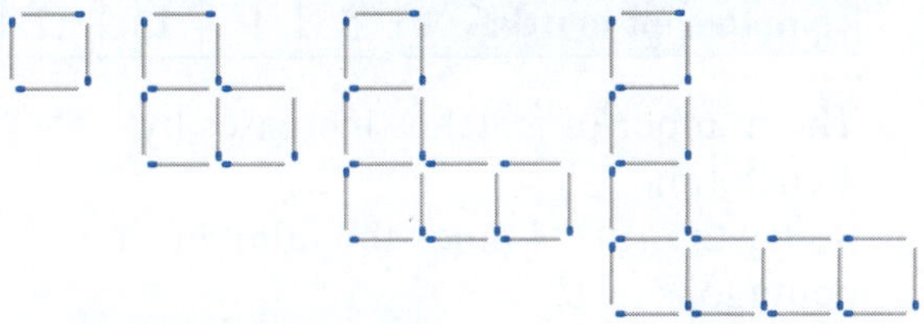

| Shape | 1 | 2 | 3 | 4 |
|---|---|---|---|---|
| Number of matches | 4 | 10 | 16 | 22 |

The pattern for the number of matches is 4, 10, 16, 22, …

The numbers are increasing by 6 each time.
The next number is $22 + 6 = 28$.
So 28 matches will be needed for Shape 5.

**43** Try each option:
$0.28 < 0.30$
$0.7 > 0.3$
$0.0465 < 0.3000$
$0.009 < 0.300$
$0.541 > 0.300$
The numbers smaller than 0.3 are 0.28, 0.0465 and 0.009.

**44**

| Day | Number of laps |
|---|---|
| Monday | 2 |
| Tuesday | 4 |
| Wednesday | 5 |
| Thursday | 6 |
| Friday | 3 |

Total laps $= 2 + 4 + 5 + 6 + 3$
$= 20$
Mean number of laps $= 20 \div 5$
$= 4$

**45** Number of pencils $= 8 \times 7 + 4$
$= 56 + 4$
$= 60$
Total number needed $= 12 \times 7$
$= 84$
Extra needed $= 84 - 60$
$= 24$

**46** 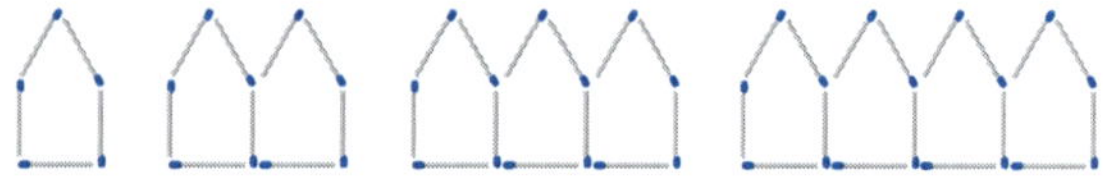

A table can be drawn up to show the number of matches needed for different numbers of pentagons.

| Number of pentagons | 1 | 2 | 3 | 4 |
|---|---|---|---|---|
| Number of matches | 5 | 9 | 13 | 17 |

The number of matches increases by 4 each time.
So try the rule '4 times the number of pentagons + 1'.
When there is 1 pentagon,
Number of matches $= 4 \times 1 + 1$
$= 4 + 1$
$= 5$ ✓
When there are 2 pentagons,
Number of matches $= 4 \times 2 + 1$
$= 8 + 1$
$= 9$ ✓
When there are 3 pentagons,
Number of matches $= 4 \times 3 + 1$
$= 12 + 1$
$= 13$ ✓
When there are 4 pentagons,
Number of matches $= 4 \times 4 + 1$
$= 16 + 1$
$= 17$ ✓
The correct rule to find the number of matches is '4 times the number of pentagons + 1'.

**47** Consider the options:

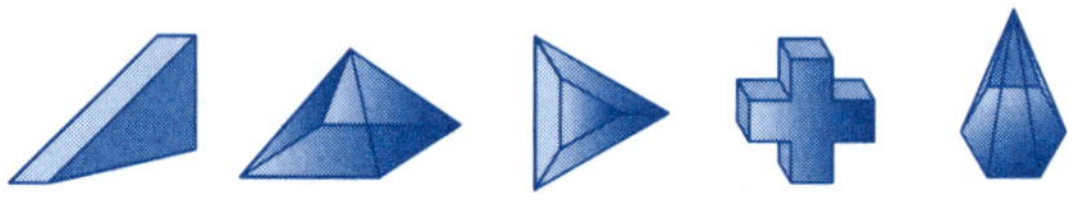

A is a triangular prism. It has 5 faces and 6 vertices. It does not have the same number of faces and vertices.
B is a rectangular pyramid. It has 5 faces and 5 vertices. It does have the same number of faces and vertices.
C is a truncated triangular prism. It has 5 faces and 6 vertices. It does not have the same number of faces and vertices.
D is a dodecagonal prism. It has 14 faces and 24 vertices. It does not have the same number of faces and vertices.
E is a hexagonal pyramid. It has 7 faces and 7 vertices. It does have the same number of faces and vertices.
The correct options are B and E.

**48** Consider each rectangle.

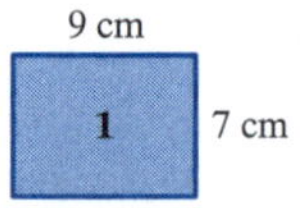


Rectangle 1:
$A = 9 \times 7$
$= 63\ \text{cm}^2$
So this is not one of the required rectangles.
Rectangle 2:
$A = 14 \times 2$
$= 28\ \text{cm}^2$
$P = 2 \times (14 + 2)$
$= 32\ \text{cm}$

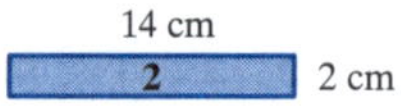


So this is the last required rectangle.
Rectangle 3:
$A = 12 \times 4$
$= 48\ \text{cm}^2$

$P = 2 \times (12 + 4)$
$= 32$ cm

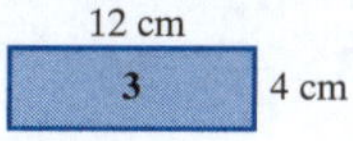


So, this is the second required rectangle.

Rectangle 4:

$A = 15 \times 3$
$= 45\ \text{cm}^2$

$P = 2 \times (15 + 3)$
$= 36$ cm

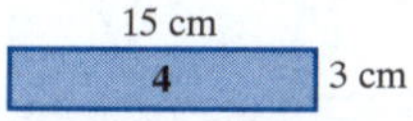


So, this is not one of the required rectangles.

Rectangle 5:

$A = 9 \times 5$
$= 45\ \text{cm}^2$

$P = 2 \times (9 + 5)$
$= 28$ cm

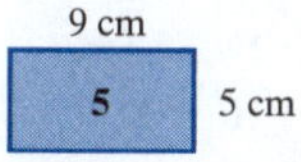


So, this is the third required rectangle.

Rectangle 6:

$A = 8 \times 6$
$= 48\ \text{cm}^2$

$P = 2 \times (8 + 6)$
$= 28$ cm

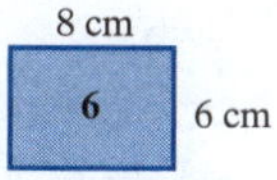


So this is the first required rectangle.

Area 48 cm$^2$
Perimeter 28 cm

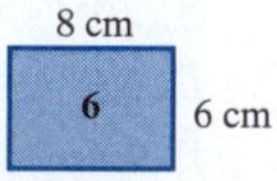


Area 48 cm$^2$
Perimeter 32 cm

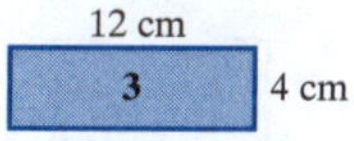


Area 45 cm$^2$
Perimeter 28 cm

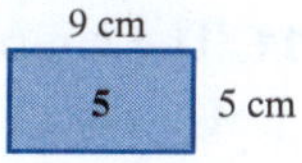


Area 28 cm$^2$
Perimeter 32 cm

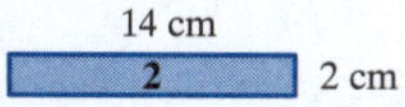


The required numbers are 6, 3, 5, 2.

## Intermediate level

### YEAR 7 Numeracy Sample Test 2

Pages 48–59

**1** 72 ha
**2** C
**3** 30%
**4** $1\frac{3}{4}$
**5** A
**6** C
**7** B, D
**8** $\frac{1}{2}, \frac{7}{12}, \frac{2}{3}, \frac{3}{4}, \frac{4}{5}, \frac{5}{6}$
**9** B
**10** D
**11** B
**12** E
**13** D
**14** B
**15** C
**16** D
**17** C
**18** C
**19** D
**20** 70 km/h
**21** 1200 m
**22** A
**23** 11
**24** C, F
**25** D, E
**26** D
**27** D
**28** −9, −5, −2.5, −2, −1.7, 0, 3, 8
**29** C
**30** 81 m$^2$
**31** 0.3, 0.26, 0.235, 0.2178, 0.08
**32** D
**33** A
**34** 9:15, 2:45, 8:20, 3:40
**35** 72 cm
**36** B, E
**37** 8
**38** B
**39** A
**40** E
**41** 2, 3, 5, 7, 11, 13, 17, 19
**42** A
**43** A
**44** $P(-3, -1)$, $Q(3, -3)$, $R(1, 0)$, $S(-1, 3)$
**45** 15 km
**46** 168
**47** B
**48** $\frac{1}{6}, \frac{1}{4}, \frac{1}{3}, \frac{1}{2}$

**1** One-sixth of the area = 12 ha
Area = $6 \times 12$ ha
= 72 ha

**2** The temperature was 9 degrees below freezing point. If it becomes 12 degrees colder it will be another 12 degrees further below freezing point.
Now $12 + 9 = 21$.
So the temperature will be 21 degrees below freezing point or $-21$ °C.

# Year 7 Numeracy Sample Online-style Test Answers

**3** The number of 13-year-old students is missing.

| Age of students | Number of students |
|---|---|
| 11 | 7 |
| 12 | 16 |
| 13 | |
| 14 | 12 |
| Total | 50 |

Total of other ages $= 7 + 16 + 12$
$= 35$
Number of 13-year-old students $= 50 - 35$
$= 15$
Fraction of 13-year-olds $= \frac{15}{50}$
$= \frac{30}{100}$
So the percentage of students who are 13 is 30%.

**4** $2\frac{1}{2} - \frac{3}{4} = 2\frac{2}{4} - \frac{3}{4}$
$= 2 - \frac{1}{4}$
$= 1\frac{3}{4}$

**5** $36 \times \square = 16$
$\square = 16 \div 36$
$= \frac{16}{36}$
$= \frac{4}{9}$

**6** Each paver is 0.35 m across.
The distance to be covered is 7.7 m.
So you need to find how many 0.35s there are in 7.7.
Required calculation $= 7.7 \div 0.35$

**7** $4 \times 9 = 36$
Now consider the options:
$2 + 10 \times 3 = 2 + 30$
$= 32$
$5 \times 8 - 4 = 40 - 4$
$= 36$
$2 \times 12 + 10 = 24 + 10$
$= 34$
$7 \times 5 + 1 = 35 + 1$
$= 36$
The expressions that equal $4 \times 9$ are $5 \times 8 - 4$ and $7 \times 5 + 1$.

**8** $\frac{30}{60} = \frac{1}{2}$ [dividing both numerator and denominator by 30]
$\frac{35}{60} = \frac{7}{12}$ [dividing both numerator and denominator by 5]
$\frac{40}{60} = \frac{2}{3}$ [dividing both numerator and denominator by 20]
$\frac{45}{60} = \frac{3}{4}$ [dividing both numerator and denominator by 15]
$\frac{48}{60} = \frac{4}{5}$ [dividing both numerator and denominator by 12]
$\frac{50}{60} = \frac{5}{6}$ [dividing both numerator and denominator by 10]

**9** Forty thousand is 40000.
$40000 + 300 + 2 = 40302$

**10** There are 13 cubes altogether.

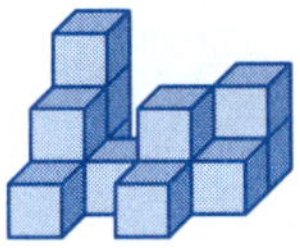

5 cubes can be seen in the top and middle rows so there can only be 8 cubes in the bottom row.
So the view from the top must be D.

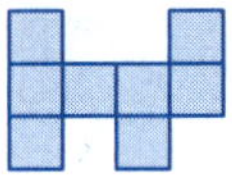

**11** The shape turns through 270° in an anticlockwise direction.

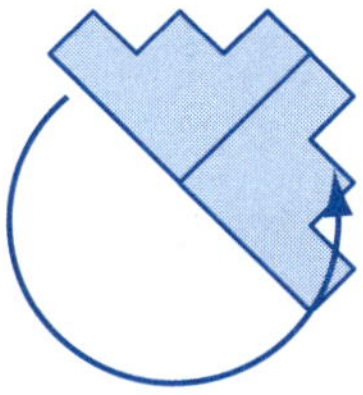

The resulting shape is B.

**12** $\frac{19.72 - 16.8}{0.16 - 0.08} = \frac{2.92}{0.08}$
$= 36.5$

**13** The scale on jug D has 2.5 litres divided into 10 parts so each part is 0.25 litres or 250 mL.
So jug D holds 2 litres of juice. No other jug has 2 litres of juice so jug D has the most.

# Year 7 Numeracy Sample Online-style Test Answers

**14** Write in some of the values on the number line.

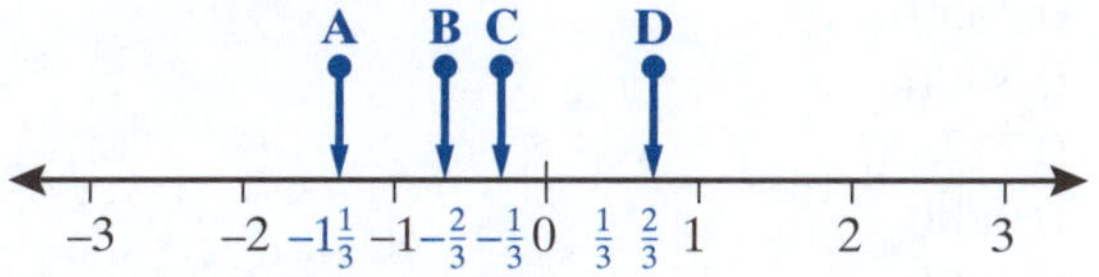


The location of $-\frac{2}{3}$ is B.

**15** Every 2 discs balance 3 blocks.
So 3 × 2 discs will balance 3 × 3 blocks.
So 6 discs balance 9 blocks.
6 discs and 1 block will be balanced by 10 blocks.

**16** $x + 60 = 180$ (angles on a straight line)

So $x = 180 - 60$

$= 120$

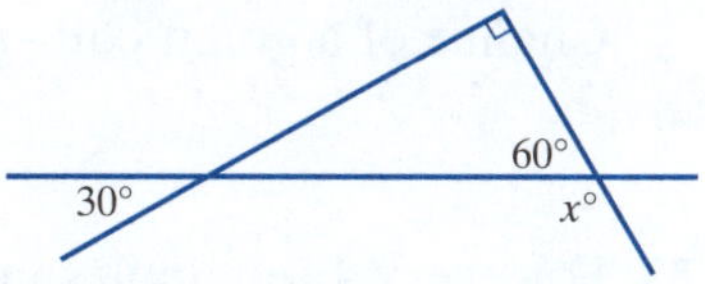


**17** Winona turns into Dean Street.

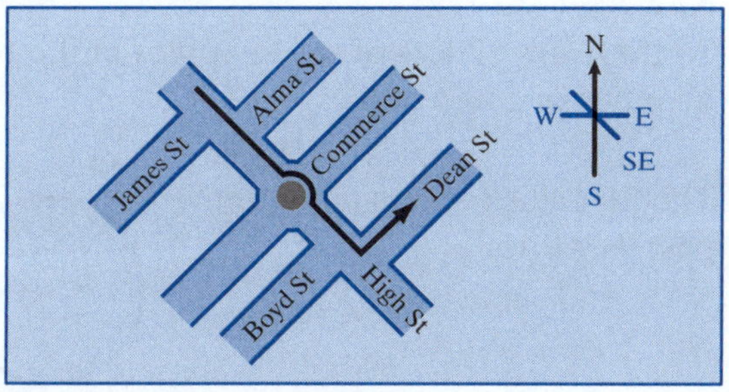


**18** 1 m and 15 cm = 100 cm + 15 cm
= 115 cm
[But this is not one of the options.]
Now 100 cm = 1 m
So 115 cm = 1.15 m
[But this is not one of the options.]
Now 1 cm = 10 mm
So 115 cm = 1150 mm
The correct option is C.

**19** The octagon has 8 lines of symmetry.

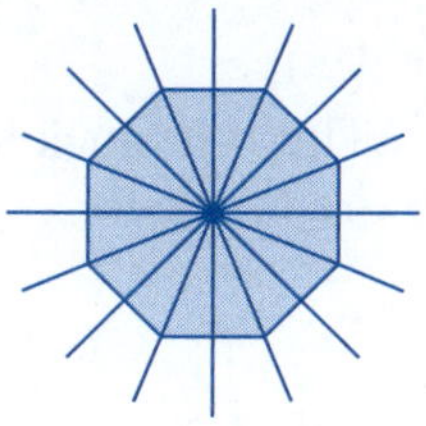

**20** Distance = 280 km
From 8:35 am until 12:35 pm is 4 hours.
Speed = 280 km ÷ 4 h
= 70 km/h

**21** On the map it is 5 squares from the gate to the house.

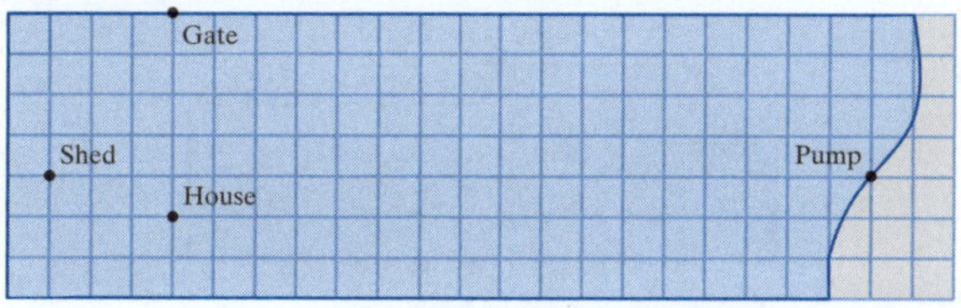


So 5 squares represent 300 m.
So 1 square represents 300 m ÷ 5 or 60 m.
There are 20 squares from the shed to the pump.
Distance = 20 × 60 m
= 1200 m

**22** Vincent's tin must be A as there is only 1 blue button out of 5 in this tin so the chance of choosing a blue button could be described as unlikely.

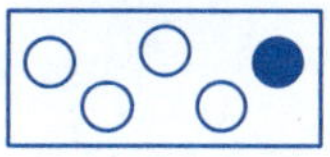

The chance of a blue button would be described as impossible in B, likely in C and certain in D.

**23** After adding 3, Amelia's answer was 47.
Before adding 3, it must have been 44.
After multiplying by 4, the answer was 44.
Before multiplying by 4 it must have been 11.
Amelia's number was 11.

**24** An obtuse angle measures more than 90° but less than 180°. The obtuse angles are C and F.

[A and B are reflex angles, D is acute and E is a right angle.]

**25** Write each decimal with the same number of places after the decimal point.
0.0900, 0.1300, 0.1040, 0.0078, 0.0250
Now 0.06 is the same as 0.0600.
The numbers that are less than 0.06 are 0.0078 and 0.025

**26** The depth of water is less than 1.75 m.
Of the options, the depth must be 1.74 m or 174 cm.

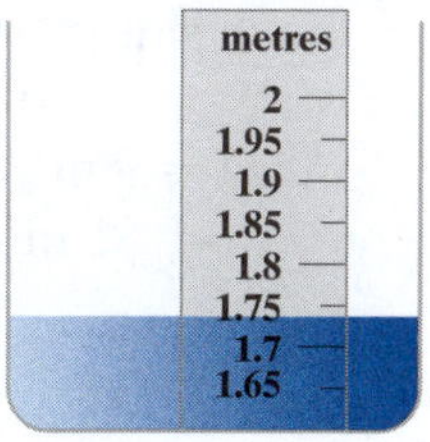

**27** Two measurements are not shown on the diagram.

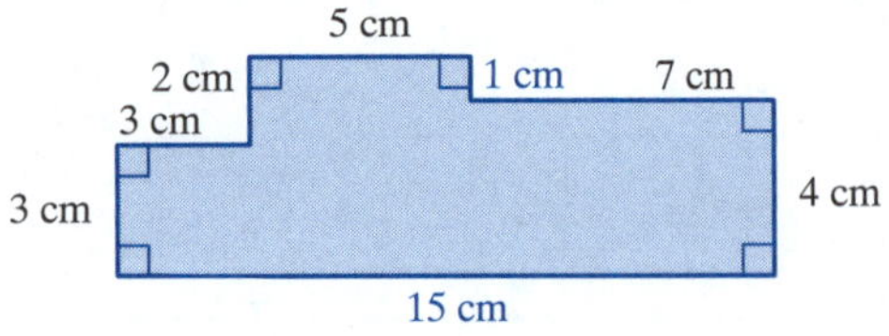


The total of the top measurements is 15 cm so the bottom must also be 15 cm. The total left side is 5 cm so the total right side must also be 5 cm.

Perimeter $= 2 \times (15 + 5)$ cm
$= 2 \times 20$ cm
$= 40$ cm

**28** Numbers less than zero are $-2.5, -5, -2, -9$ and $-1.7$. Of those $-9$ is the lowest, followed by $-5$. $-2.5$ is less than $-2$ but $-1.7$ is greater than $-2$.

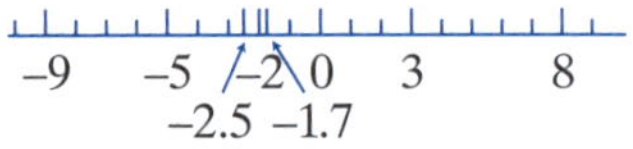


So, in order, the numbers are $-9, -5, -2.5, -2, -1.7, 0, 3, 8$.

**29** Prism I is 3 by 3 by 3.
Volume in cubic units $= 3 \times 3 \times 3$
$= 27$

Prism II is 4 by 3 by 2.
Volume in cubic units $= 4 \times 3 \times 2$
$= 24$

Prism III is 6 by 2 by 2.
Volume in cubic units $= 6 \times 2 \times 2$
$= 24$

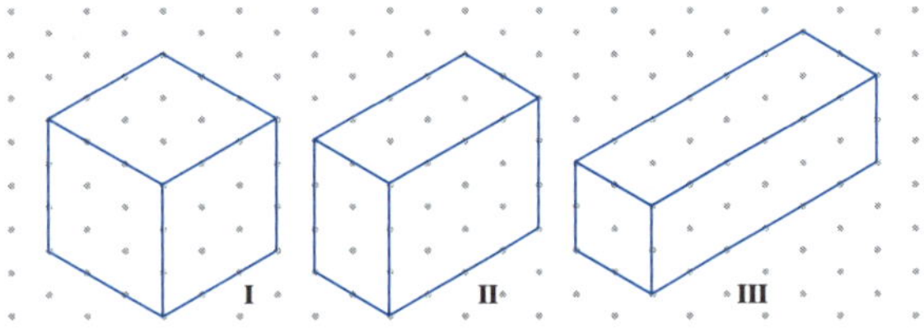


Prisms II and III have the same volume.

**30** The 4 sides of the square total 36 metres.
Each side $= 36 \text{ m} \div 4$
$= 9$ m
Area $= 9 \text{ m} \times 9 \text{ m}$
$= 81 \text{ m}^2$

**31** Write each of the numbers with the same number of places after the decimal point.
0.2350
0.3000
0.2178
0.0800
0.2600
So the highest number is 0.3000 and the lowest is 0.0800.
In order from highest to lowest the numbers are 0.3, 0.26, 0.235, 0.2178 and 0.08

**32** Number of balls $= 5 + 3 + 2$
$= 10$
2 out of 10 balls are green.

Chance of a green ball $= \dfrac{2}{10}$
$= \dfrac{1}{5}$

**33** For every 4 red apples in the box, 3 are green.
So out of every 7 apples, 3 are green.
Now $84 \div 7 = 12$
So there are 12 lots of seven apples in the box.
Number of green apples $= 12 \times 3$
$= 36$

**34** The first clock shows a quarter past 9 or 9:15.

The second clock shows a quarter to 3 or 2:45.

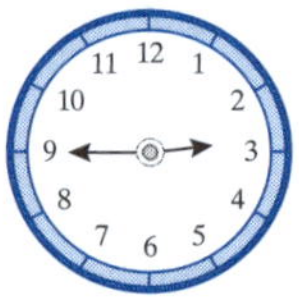

The third clock shows twenty past 8 or 8:20.

The last clock shows twenty to 4 or 3:40.

# Year 7 Numeracy Sample Online-style Test Answers

**35** Old width = 8 cm
New width = 48 cm
Number of times larger $= 48 \div 8$
$= 6$
Old length = 12 cm
New length $= 6 \times 12$ cm
$= 72$ cm

**36** 1 kilolitre = 1000 litres
2 kL + 50 L = 2000 L + 50 L
= 2050 L
= 2.05 kL
So 2050 L and 2.05 kL are both the same as 2 kilolitres and 50 litres.

**37** The graph represents 100 cars. The graph is divided into 25 sections so each section represents 4 cars.

Number of white cars $= 6 \times 4$
$= 24$
Number of red cars $= 4 \times 4$
$= 16$
Difference $= 24 - 16$
$= 8$
There were 8 more white cars than red cars.

**38** 15 minutes is a quarter of an hour.
A quarter of an hour is 0.25 h.
So 3 h and 15 min = 3.25 h

**39** A square-based pyramid has 5 faces and 8 edges.

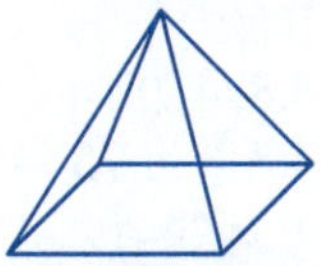

The new object has 8 faces and 12 edges.

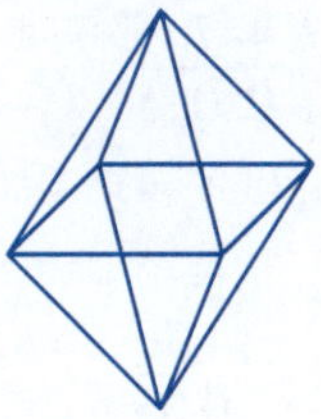

So the new object has 3 more faces and 4 more edges.

**40** The average of 6 numbers is 4.
Sum of the 6 numbers $= 6 \times 4$
$= 24$
The average of 5 numbers is 3.
Sum of the 5 numbers $= 5 \times 3$
$= 15$
Difference in sums $= 24 - 15$
$= 9$
So the number that was left out was 9.

**41** A prime number has exactly 2 factors: 1 and itself.
The prime numbers are 2, 3, 5, 7, 11, 13, 17 and 19.

**42** Average from 3 tests = 79
Total marks from 3 tests $= 3 \times 79$
$= 237$
Total marks from 4 tests $= 237 + 87$
$= 324$
Average of 4 tests $= 324 \div 4$
$= 81$

**43** The first number = 78
The second number $= 78 \div 2 - 1$
$= 39 - 1$
$= 38$
The third number $= 38 \div 2 - 1$
$= 19 - 1$
$= 18$
The fourth number $= 18 \div 2 - 1$
$= 9 - 1$
$= 8$
The fifth number $= 8 \div 2 - 1$
$= 4 - 1$
$= 3$

**44**

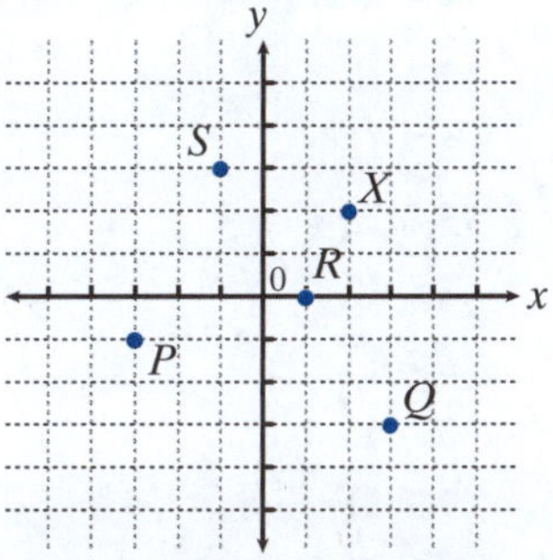


P is the point $(-3, -1)$.
Q is $(3, -3)$.
$R(1, 0)$
$S(-1, 3)$

**45** 180 km/h means 180 km in 60 minutes.
This is 18 km in 6 minutes or 3 km per minute.
In 5 minutes, Leanne will travel $5 \times 3$ km or 15 km.

**46** 60% of 700 students are boys.
Number of boys $= \frac{60}{100} \times 700$
$= 420$
40% of the boys are in Year 12.
Number of boys in Year 12 $= \frac{40}{100} \times 420$
$= 168$

**47** 750 km = 7.5 × 100 km
Amount of fuel = 7.5 × 8 L
= 60 L

**48** There are 4 steps from $\frac{1}{12}$ to $\frac{5}{12}$.
So the rule for the pattern is to add $\frac{1}{12}$ each time.
The pattern is $\frac{1}{12}, \frac{2}{12}, \frac{3}{12}, \frac{4}{12}, \frac{5}{12}, \frac{6}{12}$
or $\frac{1}{12}, \frac{1}{6}, \frac{1}{4}, \frac{1}{3}, \frac{5}{12}, \frac{1}{2}$.

## Intermediate level

### YEAR 7 Numeracy Sample Test 3

Pages 60–71

**1** B
**2** 61
**3** 600 kg
**4** D
**5** D
**6** D
**7** A, C, D
**8** $\frac{13}{20}, \frac{2}{3}, \frac{7}{10}, \frac{18}{25}, \frac{3}{4}, \frac{4}{5}$
**9** C
**10** 22 °C
**11** C
**12** E
**13** 300 028
**14** B
**15** A
**16** D
**17** C
**18** B
**19** A
**20** C
**21** C
**22** 3 L
**23** A
**24** A
**25** 0.0087, 0.06, 0.123, 0.15
**26** E
**27** E
**28** 900 mL
**29** 31
**30** C, F
**31** 550 g
**32** B
**33** C
**34** 294 cm²
**35** B
**36** C, D
**37** 19
**38** D
**39** C, D
**40** D
**41** 1, 3, 5, 4, 2
**42** 3
**43** 127°
**44** 32 m
**45** 1686
**46** 100
**47** B
**48** $3.25, $3.15, $2.80, $2.55

**1** Saving = 20% of $800
Now 10% of $800 is $80.
So 20% of $800 = 2 × $80
= $160
Price paid = $800 − $160
= $640

**2** There were 250 children altogether.

| | Walked | Did not walk |
|---|---|---|
| Boys | 73 | 52 |
| Girls | | 64 |

Number in table = 73 + 52 + 64
= 189
Number remaining = 250 − 189
= 61
The number of girls who walked is 61.

**3** One-fifth is 120 kg.
The whole = 5 × 120 kg
= 600 kg

**4** 12 + 8 ÷ 4 − 2 × 3
= 12 + 2 − 6
= 8

**5** There are 7 chairs in the first row.
There is one extra chair in every new row.
So, in 7 rows, the number of chairs
= 7 + 8 + 9 + 10 + 11 + 12 + 13
= (7 + 13) + (8 + 12) + (9 + 11) + 10
= 20 + 20 + 20 + 10
= 70

**6** 4.8, 4.75, 4.7, 4.65, …
4.8 − 4.75 = 0.05

$$\begin{array}{r} 4.{}^{7}\!\not{8}{}^{1}0 \\ -\ 4.7\ 5 \\ \hline 0.0\ 5 \end{array}$$

4.75 − 4.7 = 0.05

$$\begin{array}{r} 4.75 \\ -\ 4.70 \\ \hline 0.05 \end{array}$$

4.7 − 4.65 = 0.05
The rule is to decrease by 0.05.

**7** Consider each option:
3 − 4 × 2 + 6 = 3 − 8 + 6
= 1
12 ÷ 6 × 2 = 2 × 2
= 4
(17 + 18) ÷ (7 × 5) = 35 ÷ 35
= 1

$4 \times (21 - 12) \div 6^2 = 4 \times 9 \div 36$
$= 36 \div 36$
$= 1$

$-8 + 2 - 5 + 6 + 4 = -1$

So the number sentences $3 - 4 \times 2 + 6$, $(17 + 18) \div (7 \times 5)$ and $4 \times (21 - 12) \div 6^2$ are equal to 1.

**8** $65\% = \frac{65}{100} = \frac{13}{20}$

$66\frac{2}{3}\% = \frac{2}{3}$

$70\% = \frac{70}{100} = \frac{7}{10}$

$72\% = \frac{72}{100} = \frac{18}{25}$

$75\% = \frac{3}{4}$

$80\% = \frac{8}{10} = \frac{4}{5}$

**9** The object has 20 edges.

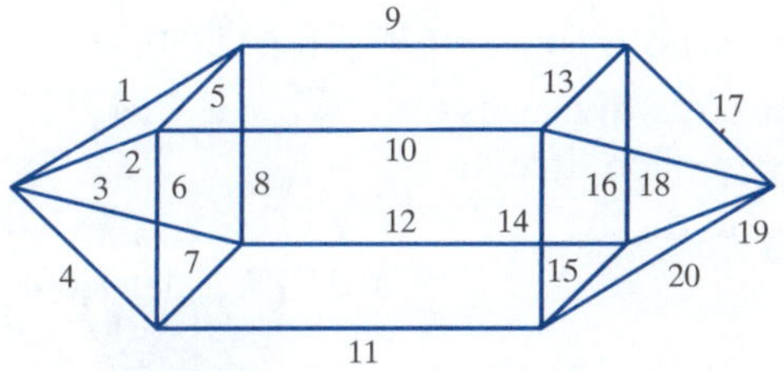


**10** Difference $= 3 - (-19)$
$= 3 + 19$
$= 22$

The difference between these temperatures is 22 °C.

**11** Travis is travelling South-East.

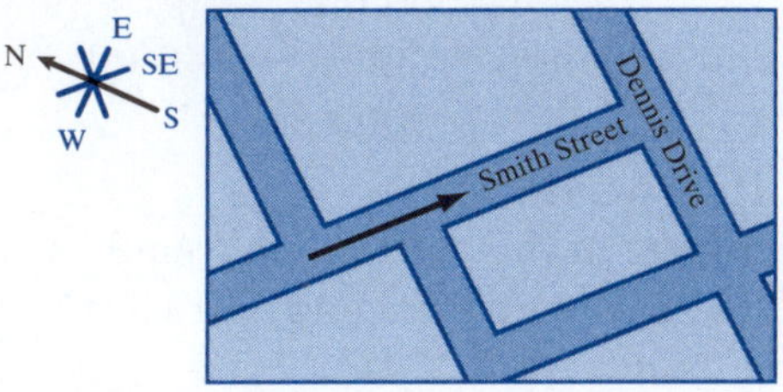


**12** 3582517 is between 3582000 and 3583000. It is closer to 3583000.
So the population is 3583000 to the nearest thousand.

**13** Three hundred thousand and twenty-eight
$= 300\,000 + 28$
$= 300\,028$

**14** An angle of 90° is a right angle and an angle of 180° is a straight angle. An angle of 120° is larger than a right angle but smaller than a straight angle. It is nearer in size to a right angle than a straight angle. The angle that measures closest to 120° is B.

**15** The number is between 0 and −0.5.
Of the options, it can only be −0.4.

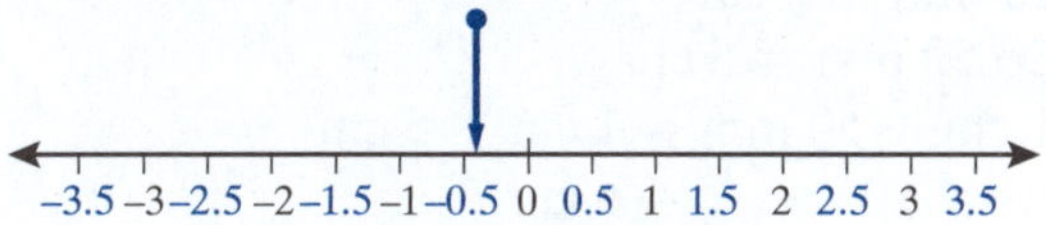


**16** The top is 15 metres so the total of the bottom (horizontal) sides is also 15 metres. The left side is 8 metres so the right side is also 8 metres. There is an extra 2 metres on the centre bottom right side so there will be an extra 2 metres on the centre bottom left side as well.

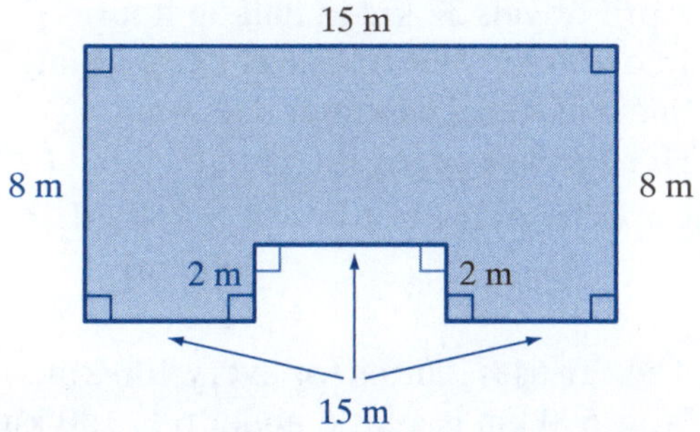


Perimeter $= 2 \times (15 + 8 + 2)$ m
$= 2 \times 25$ m
$= 50$ m

**17** Total buttons $= 25 + 12 + 15 + 8$
$= 60$

12 of the buttons are blue.

Chance of blue button $= \frac{12}{60}$
$= \frac{1}{5}$

# Year 7 Numeracy Sample Online-style Test Answers

**18** From the front you can see that the object is 4 blocks wide. The blocks are 2 high on the left and 3 high, second from the right.

When viewed from the back the position of these blocks is seen in reverse.
The shape is 4 blocks across, 2 high on the right and 3 high on the second from the left.

The correct option is B.

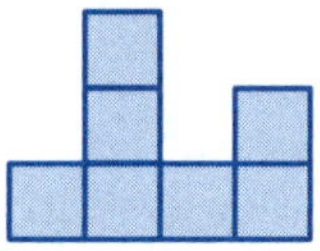

**19** 10 mm = 1 cm
So 50 mm = 5 cm
1 cm + 50 mm = 1 cm + 5 cm
= 6 cm

**20** The shapes are rotating this way

This is clockwise.
The shape makes two rotations in a quarter turn.
Angle of rotation $= 90° \div 2$
$= 45°$
The shape is being rotated 45° clockwise.

**21** April travels 32 km in half an hour.
In one hour she will travel twice as far.
She will travel 64 km in one hour.
Her average speed is 64 km/h.

**22** Capacity of the bottle $= 4 \times 750$ mL
= 3000 mL
= 3 L

**23** The car uses 7 litres for every 100 km.
Now 590 km is a little under $6 \times 100$ km so the fuel used will be a little less than $6 \times 7$ litres.
As $6 \times 7 = 42$, the best estimate will be 40 litres.

**24** The top view shows that the object is 5 blocks one way and 4 the other. The front view shows that the shape is 4 blocks across so it must be 5 blocks deep. The view from a side must show 5 blocks across, not 4. A side view cannot be A.
[The other options could be as shown below.]

B

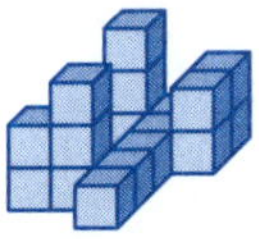

viewed from right

C

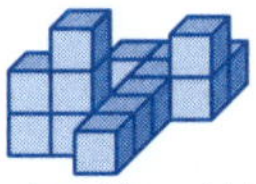

viewed from right

D

viewed from left

**25** Write each decimal with the same number of places after the decimal point.
0.0600, 0.1500, 0.0087, 0.1230
So, in order, from lowest to highest, the numbers are 0.0087, 0.0600, 0.1230 and 0.1500.
The numbers are 0.0087, 0.06, 0.123 and 0.15 in order.

**26** Each of the two pieces has 4 sides so each is a quadrilateral. Each has one pair of opposite sides that are parallel. Each of the shapes is a trapezium.

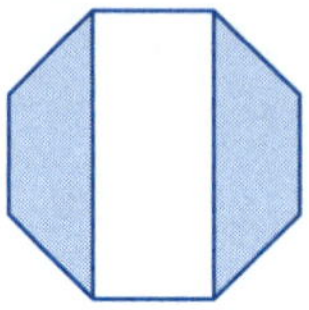

**27** The movie finishes one hour and 50 minutes after 8:40 pm.
One hour after 8:40 pm is 9:40 pm.
20 minutes after 9:40 pm is 10 pm.
Another 30 minutes after 10 pm is 10:30 pm.
Now 12 + 10 = 22.
So, in 24-hour time, 10:30 pm is 2230.

**28** There are 5 divisions between each litre in the scale on the jug.

Each division $= 1 \text{ L} \div 5$
$= 1000 \text{ mL} \div 5$
= 200 mL
So the jug held 1400 mL of juice.
Amount removed $= 2 \times 250$ mL
= 500 mL
Amount remaining = (1400 − 500) mL
= 900 mL

**29** 1, 3, 7, 15, **?**, 63, 127, 255
The numbers in the pattern are found by the rule 'Double the number and add 1'.
Missing number $= 2 \times 15 + 1$
$= 30 + 1$
$= 31$
[Check: Next number $= 2 \times 31 + 1$
$= 62 + 1$
$= 63$ ✓]
The missing number is 31.

# Year 7 Numeracy Sample Online-style Test Answers

**30** Consider each option.
In Option A the chance of landing on 4 is 1 in 3.
In Option B the chance of landing on 4 is 1 in 6.
In Option C the chance of landing on 4 is 2 in 8 or 1 in 4.
In Option D the chance of landing on 4 is 4 in 12 or 1 in 3.
In Option E, the outcomes are not equally likely. The chance of landing on 4 is less than 1 in 4.
In Option F, the change of landing on 4 is 3 in 12 or 1 in 4.
The spinners that have a 1 in 4 chance of landing on 4 are C and F.

**31** 1 kg = 1000 g
So 1.3 kg = 1300 g
Amount of meat remaining = (1300 − 750) g
= 550 g

**32** 1 cm represents 25 km.
3.6 cm represents 3.6 × 25 km or 90 km.

**33** Angles at a point add to 360°.
Now 105° + 71° + 94° = 270°
Remaining angle = 360° − 270°
= 90°

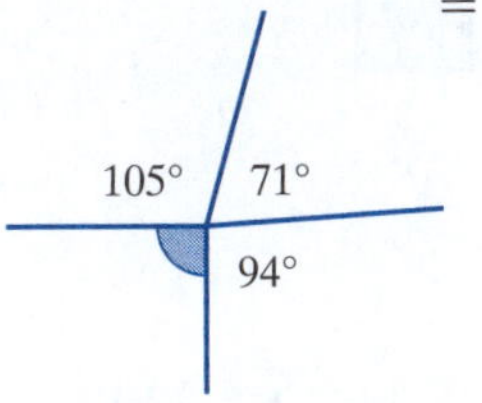


So the shaded angle must be a right angle.

**34** Each face is a square.
Area of each face = 7 cm × 7 cm
= 49 $cm^2$
The cube has 6 faces.
Total area = 6 × 49 $cm^2$
= 294 $cm^2$

**35** The clock shows a quarter to twelve or fifteen minutes to twelve.
Julia must wait 15 minutes to twelve o'clock and 10 minutes after twelve o'clock.
Waiting time = (15 + 10) min
= 25 min

**36** Consider the options:

| □ | 2 | 3 | 5 | 8 |
|---|---|---|---|---|
| △ | 0 | 5 | 21 | 60 |

△ = 5 × □ − 10
When □ = 2,
△ = 5 × 2 − 10
= 10 − 10
= 0 ✓
When □ = 3,
△ = 5 × 3 − 10
= 15 − 10
= 5 ✓
When □ = 5,
△ = 5 × 5 − 10
= 25 − 10
= 15 ✗
The rule is not △ = 5 × □ − 10.

△ = 3 × □ − 6
When □ = 2,
△ = 3 × 2 − 6
= 6 − 6
= 0 ✓
When □ = 3,
△ = 3 × 3 − 6
= 9 − 6
= 3 ✗
The rule is not △ = 3 × □ − 6.

△ = □ × □ − 4
When □ = 2
△ = 2 × 2 − 4
= 4 − 4
= 0 ✓
When □ = 3,
△ = 3 × 3 − 4
= 9 − 4
= 5 ✓
When □ = 5,
△ = 5 × 5 − 4
= 25 − 4
= 21 ✓
When □ = 8,
△ = 8 × 8 − 4
= 64 − 4
= 60 ✓
The rule could be △ = □ × □ − 4

$\triangle = (\square + 2) \times (\square - 2)$
When $\square = 2$
$\triangle = (2 + 2) \times (2 - 2)$
$= 4 \times 0$
$= 0$ ✓
When $\square = 3$,
$\triangle = (3 + 2) \times (3 - 2)$
$= 5 \times 1$
$= 5$ ✓
When $\square = 5$,
$\triangle = (5 + 2) \times (5 - 2)$
$= 7 \times 3$
$= 21$ ✓
When $\square = 8$,
$\triangle = (8 + 2) \times (8 - 2)$
$= 10 \times 6$
$= 60$ ✓
The rule could be $\triangle = (\square + 2) \times (\square - 2)$.
So the rule could be either $\triangle = \square \times \square - 4$ or $\triangle = (\square + 2) \times (\square - 2)$.

**37** In Year 7 there are 32 boys and 38 girls. There are 6 more girls than boys.

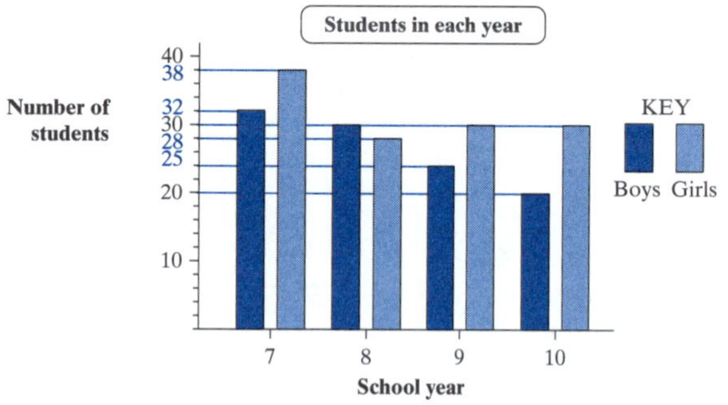


In Year 8 there are 30 boys and 28 girls. There are 2 fewer girls than boys.
In Year 9 there are 25 boys and 30 girls. There are 5 more girls than boys.
In Year 10 there are 20 boys and 30 girls. There are 10 more girls than boys.
Total of extra girls $= 6 - 2 + 5 + 10$
$= 19$
There are 19 more girls than boys.

**38** The ratio of men to women is 3 to 5 so out of every 8 adults, 5 are women.
Now $560 \div 8 = 70$
So there are 70 lots of 8 adults.
Number of women $= 5 \times 70$
$= 350$

**39** The opposite faces must add to 7.
This means that Face 1 must be opposite Face 6, Face 2 opposite Face 5 and Face 3 opposite Face 4.
This occurs with Options C and D.
[Option A has 1 opposite 4, 2 opposite 5 and 3 opposite 6. Option B has 1 opposite 4, 2 opposite 6 and 3 opposite 5. Option E has 1 opposite 6, 2 opposite 3 and 4 opposite 5.]

**40** \$4500 is 20%.
The whole amount is 100%.
Now $20\% \times 5 = 100\%$
So the whole amount $= 5 \times \$4500$
$= \$22\,500$

**41** The area of any rectangle is found by multiplying the length by the breadth.
$8 \text{ cm} \times 4 \text{ cm} = 32 \text{ cm}^2$
$6 \text{ cm} \times 6 \text{ cm} = 36 \text{ cm}^2$
$11 \text{ cm} \times 3 \text{ cm} = 33 \text{ cm}^2$
$5 \text{ cm} \times 7 \text{ cm} = 35 \text{ cm}^2$
$17 \text{ cm} \times 2 \text{ cm} = 34 \text{ cm}^2$
So, in order from smallest area to largest area, the rectangles are:

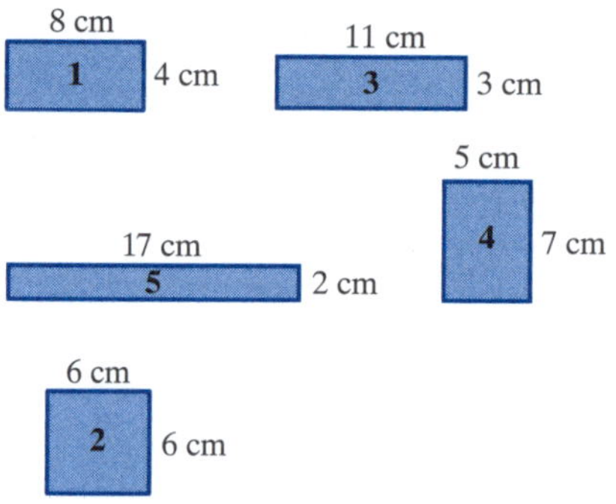


The required numbers are 1, 3, 5, 4 and 2.

**42** After 7 was added the answer was 16.
Before 7 was added the answer was $16 - 7$ or 9.
Now $3 \times 3 = 9$
So the number was 3.

**43** Vertically opposite angles are equal.
So shaded angle $= 90° + 37°$
$= 127°$

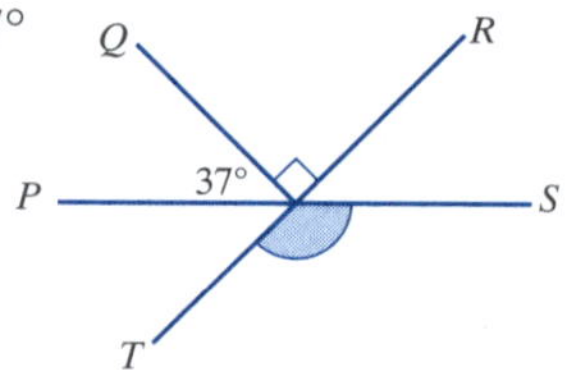


**44** Area $= 64 \text{ m}^2$
Now $64 = 8 \times 8$
So each side of the square is 8 m.
Perimeter $= 4 \times 8 \text{ m}$
$= 32 \text{ m}$

**45** First number = 1

Second number = 1 × 1 + 5
= 1 + 5
= 6

Third number = 6 × 6 + 5
= 36 + 5
= 41

Fourth number = 41 × 41 + 5
= 1681 + 5
= 1686

**46** Total marks after 6 tests = 6 × 65
= 390

To have an average of 70 after 7 tests:

Total marks = 7 × 70
= 490

Difference = 490 − 390
= 100

Jasmine would need to score 100 in her seventh test to have an average of 70.

**47** Halfway between $3\frac{3}{4}$ and $5\frac{1}{4}$ is halfway between 4 and 5 (after adding $\frac{1}{4}$ to $3\frac{3}{4}$ and subtracting $\frac{1}{4}$ from $5\frac{1}{4}$).

Halfway between 4 and 5 is $4\frac{1}{2}$.

**48** Price = 2.5 × $1.30
= $3.25

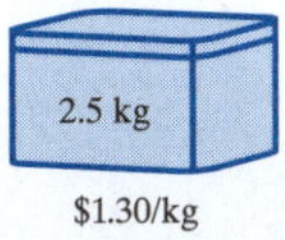


Price = 0.75 × $4.20
= $3.15

Price = 1.6 × $1.75
= $2.80

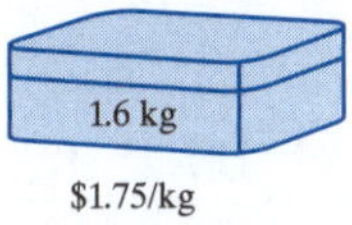


Price = 4.25 × $0.60
= $2.55

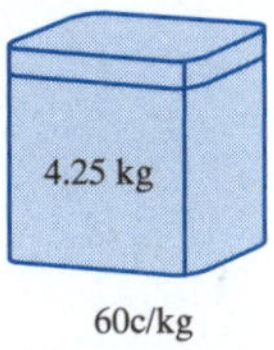


## Advanced level

### YEAR 7 Numeracy Sample Test 4

Pages 72–85

| | | | |
|---|---|---|---|
| **1** | 7 | **24** | D |
| **2** | C | **25** | 120 $cm^3$ |
| **3** | C, E | **26** | B |
| **4** | $\frac{1}{2}$ | **27** | 875 km |
| **5** | D | **28** | D |
| **6** | E | **29** | 1.9 km |
| **7** | 0.75, 0.8, $0.\dot{6}$, 0.7, 0.85 | **30** | 6 mL |
| **8** | $2.00/L, $1.75/L, $1.65/L, $1.70/L | **31** | A, B, C |
| **9** | D | **32** | A |
| **10** | 28 | **33** | B, D |
| **11** | 1650 mL | **34** | A |
| **12** | 4.8 cm | **35** | A |
| **13** | 81 | **36** | B |
| **14** | C | **37** | C |
| **15** | A | **38** | 2.4 L |
| **16** | B | **39** | B |
| **17** | B | **40** | C |
| **18** | 120° | **41** | C, D |
| **19** | 1.4 | **42** | $\frac{7}{24}, \frac{1}{3}, \frac{3}{8}, \frac{5}{12}$ |
| **20** | C | **43** | A |
| **21** | B | **44** | C |
| **22** | A | **45** | 91 |
| **23** | 12 km | **46** | 110 |
| | | **47** | A, C, D |
| | | **48** | 350 km |

**1** 10, 9.25, 8.5, 7.75, …

10 − 9.25 = 0.75

$$\begin{array}{r} \not{1}{}^{9}{}^{1}\not{0}.{}^{9}{}^{1}\not{0}{}^{1}0 \\ -\;\;9.\,2\,5 \\ \hline 0.\,7\,5 \end{array}$$

9.25 − 8.5 = 0.75

$$\begin{array}{r} {}^{8}\not{9}.{}^{1}25 \\ -\;\;8.\,50 \\ \hline 0.\,75 \end{array}$$

The numbers are decreasing by 0.75 each time.

The next number = 7.75 − 0.75
= 7

# Year 7 Numeracy Sample Online-style Test Answers

**2** One-third of 600 people are competitors.

Number of competitors $= 600 \div 3$
$= 200$

Number of spectators $= 600 - 200$
$= 400$

Number of spectators who are children
$= 40\%$ of 400
$= \frac{40}{100} \times 400$
$= 160$

**3** The area could be found by finding the area of the whole rectangle minus the missing rectangle.

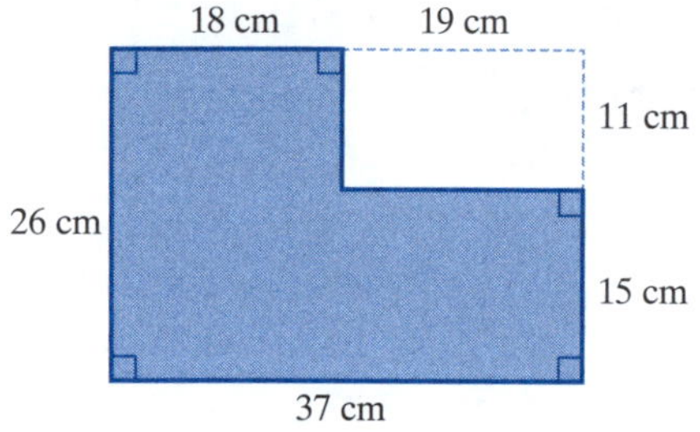


Area $= (37 \times 26 - 19 \times 11)$ cm$^2$

[This is the same as $(37 \times 26 - 11 \times 19)$ cm$^2$.]

The area could be found by dividing the shape into 2 rectangles using a horizontal line.

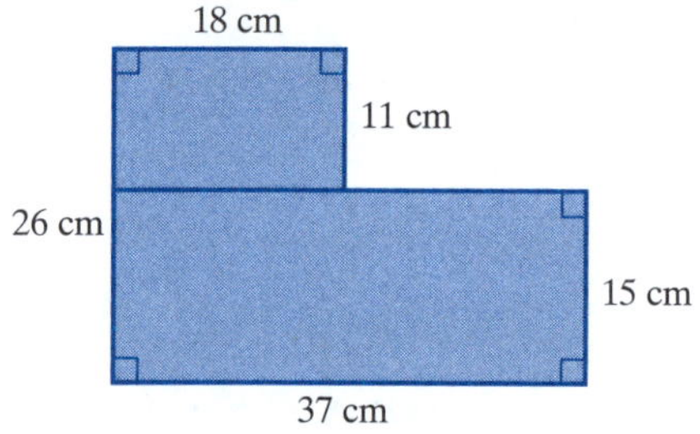


Area $= (37 \times 15 + 18 \times 11)$ cm$^2$

[But this is not one of the options.]

The area could be found by dividing the shape into 2 rectangles using a vertical line.

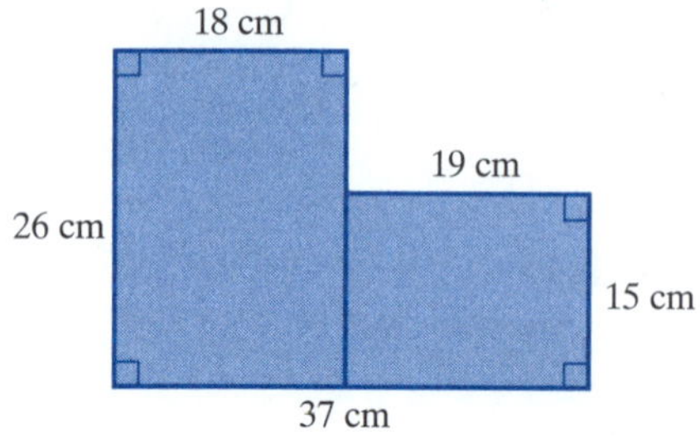


Area $= (26 \times 18 + 19 \times 15)$ cm$^2$

The number sentences $37 \times 26 - 11 \times 19$ and $26 \times 18 + 19 \times 15$ will give the shaded area.

**4** $\frac{2}{5} = \frac{4}{10}$

and $\frac{3}{5} = \frac{6}{10}$

Number halfway between $\frac{4}{10}$ and $\frac{6}{10} = \frac{5}{10}$
$= \frac{1}{2}$

So the number halfway between $\frac{2}{5}$ and $\frac{3}{5}$ is $\frac{1}{2}$.

**5** $\frac{1}{12}, \frac{1}{6}, \frac{1}{4}, \frac{1}{3}, \ldots$

[Change all the fractions to equivalent fractions with the same denominator, 12.]

$\frac{1}{6} = \frac{2}{12}$ [after multiplying numerator and denominator by 2]

$\frac{1}{4} = \frac{3}{12}$ [after multiplying numerator and denominator by 3]

$\frac{1}{3} = \frac{4}{12}$ [after multiplying numerator and denominator by 4]

$\frac{1}{12}, \frac{2}{12}, \frac{3}{12}, \frac{4}{12}, \ldots$

The next number in the pattern is $\frac{5}{12}$.

**6** Fraction on area and volume $= \frac{2}{5} + \frac{1}{3}$
$= \frac{6}{15} + \frac{5}{15}$
$= \frac{11}{15}$

Fraction on perimeter $= 1 - \frac{11}{15}$
$= \frac{4}{15}$

**7** $\frac{3}{4} = 0.75$

$\frac{4}{5} = \frac{8}{10} = 0.8$

$\frac{2}{3} = 0.\dot{6}$

$\frac{7}{10} = 0.7$

$\frac{17}{20} = \frac{85}{100} = 0.85$

**8** [Find the price for one litre for each option.]
750 mL for $1.50
250 mL for $0.50 [after dividing by 3]
1 L for $2.00 [after multiplying by 4]
So the cost is $2.00/L.

$1.50

2 L for $3.50
1 L for $1.75 [after dividing by 2]
So the cost is $1.75/L.

$3.50

3 L for $4.95
1 L for $1.65 [after dividing by 3]
So the cost is $1.65/L.

$4.95

5 L for $8.50
1 L for $1.70 [after dividing by 5]
So the cost is $1.70/L.

$8.50

**9** Isabel was facing West. A turn through 180° would mean she is facing East. Another 45° in a clockwise direction would mean she is facing South-East.

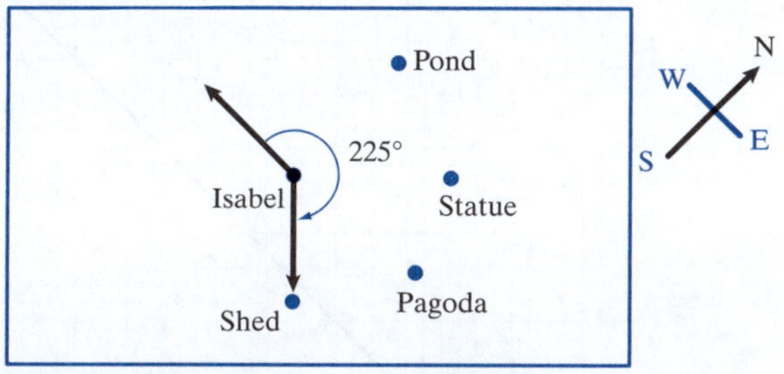


The feature that Isabel is facing is the shed.

**10** After multiplying by 5, the answer was 185.
Before multiplying by 5, the number was 185 ÷ 5 or 37.
Before adding 9, the number was 37 − 9 or 28.

**11** Each litre is divided into 4.
So each division on the scale is one-quarter of a litre (0.25 L) or 250 mL.
So the amount of milk in the jug is 2 litres and 250 mL or 2250 mL.

After 600 mL is removed:
Amount in jug = (2250 − 600) mL
= 1650 mL

**12** 1 cm represents 15 km.
From Clifton to Underwood = 72 km
[To find the distance on the map you need to know how many lots of 15 km there are in 72 km.]
Distance on map = (72 ÷ 15) cm
= 4.8 cm

**13** $\frac{2}{3}$ of a number is 72.
$\frac{1}{3}$ of the number = 72 ÷ 2
= 36
The number = 36 × 3
= 108
$\frac{1}{4}$ of the number = 108 ÷ 4
= 27
$\frac{3}{4}$ of the number = 3 × 27
= 81

**14** Number on committee = 2 + 3 + 4 + 6
= 15
The number of students from Year 8 is 3.
The chance that the student will be from Year 8 is 3 in 15 or 1 in 5.

**15** If Lydia sells 10 necklaces the amount she receives is 10 × $8 or $80.
Cost to produce 10 necklaces:
Cost = 5 × 10 + 25
= 50 + 25
= 75
The cost to make the necklaces is $75.
Profit = $80 − $75
= $5

**16** From 2:30 pm one day until 8:30 am the next day is 18 hours.
So Brisbane time is 18 hours ahead of Los Angeles time.
18 hours after 8:30 am Monday is 2:30 am Tuesday.

**17** The irrigator takes 9 hours to move 240 paces.
So it takes 3 hours to move 80 paces and
6 hours to move 160 paces.
6 hours after 10 pm is 4 am.

**18** A complete revolution is 360°.
The clockface is divided into 12 sections.
Each section $= 360° \div 12$
$= 30°$
At 4 o'clock, there are 4 sections between the hands.
Obtuse angle $= 4 \times 30°$
$= 120°$

**19** $9.23 - 4.65 = \square + 3.18$
Now $9.23 - 4.65 = 4.58$
So $\square + 3.18 = 4.58$
$\square = 4.58 - 3.18$
$= 1.4$

**20** There are two octagonal faces, both with area 120 cm$^2$.

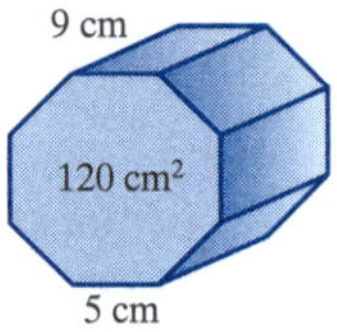


There are 8 rectangular faces.
Area of each rectangle $= 9 \text{ cm} \times 5 \text{ cm}$
$= 45 \text{ cm}^2$
Total area $= (2 \times 120 + 8 \times 45) \text{ cm}^2$
$= (240 + 360) \text{ cm}^2$
$= 600 \text{ cm}^2$

**21** 16 of 60 ribbons are pink.
Consider the options:
$\frac{1}{3}$ of $60 = 60 \div 3$
$= 20$
$\frac{1}{4}$ of $60 = 60 \div 4$
$= 15$
$\frac{1}{5}$ of $60 = 60 \div 5$
$= 12$
$\frac{1}{6}$ of $60 = 60 \div 6$
$= 10$
Of the choices, 16 is closest to 15. So the fraction of pink ribbons is closest to $\frac{1}{4}$.

**22** From the right side view, the blocks are just one high in the front row.
The only possible option is A.

**23** 45 minutes is $\frac{3}{4}$ of an hour.
In one hour the cyclist travels 16 km.
So in $\frac{3}{4}$ h the cyclist travels $\frac{3}{4}$ of 16 km or 12 km.
The distance from Eaglevale to Sandy Beach is 12 km.

**24** Each angle of a square = 90°.
Each angle of an equilateral triangle = 60°.

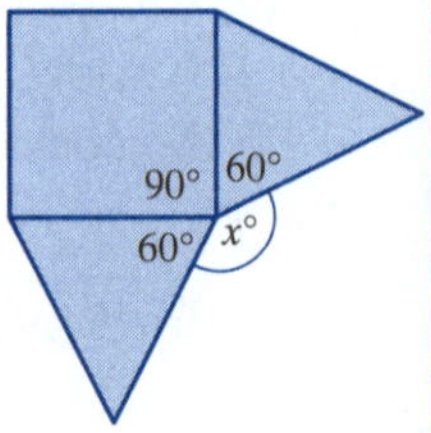


Angles at a point add to 360°.
So $x + 60 + 90 + 60 = 360$
$x + 210 = 360$
$x = 360 - 210$
$x = 150$

**25** Volume of first prism = 240 cm$^3$
If the length stayed the same while the width halved and the height doubled, the volume would still be the same.
As the length also halved, the volume will have halved.
The new volume is 120 cm$^3$.

**26** 3 hectares is about 7.5 acres.

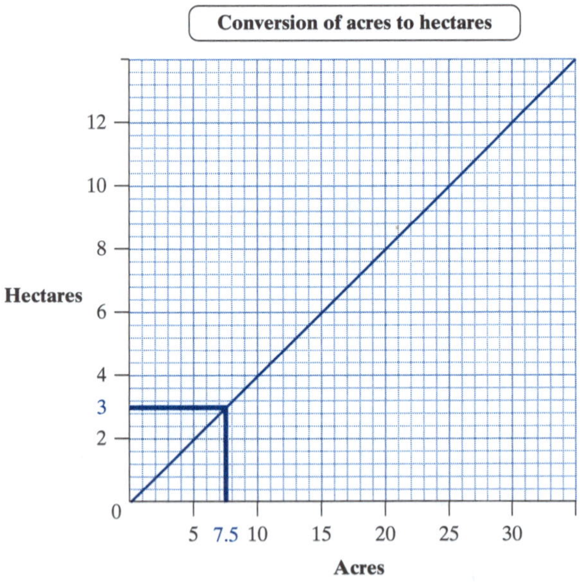


For each acre, 50 kg of fertiliser is required.
Amount of fertiliser $= 7.5 \times 50$ kg
$= 375$ kg

# Year 7 Numeracy Sample Online-style Test Answers

**27** Dane's car uses 8 L/100 km.
So it uses 1 litre for every (100 ÷ 8) km.
It uses 1 litre for every 12.5 km.
On 70 litres it will travel (70 × 12.5) km or 875 km.

**28** The average of 8 numbers is 12.
Sum of the numbers = 8 × 12
= 96
The average of 9 numbers is 14.
Sum of those numbers = 9 × 14
= 126
Difference = 126 − 96
= 30
The number that has been included must be 30.

**29**

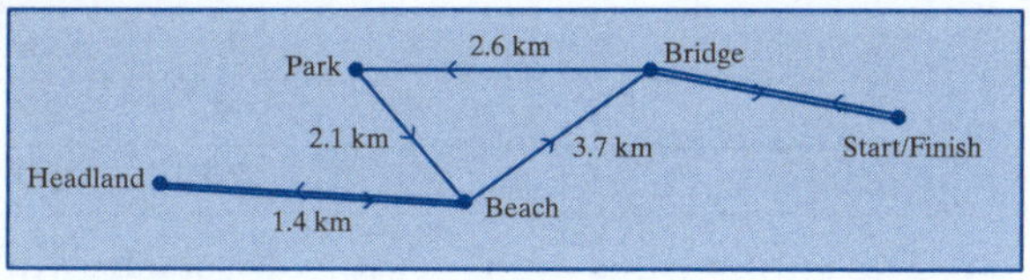


The path from the headland to the beach is covered twice.
Total distance shown
= (2 × 1.4 + 2.1 + 2.6 + 3.7) km
= (2.8 + 2.1 + 2.6 + 3.7) km
= 11.2 km
Remaining distance = (15 − 11.2) km
= 3.8 km
The distance from the start/finish to the bridge is also covered twice.
This distance = (3.8 ÷ 2) km
= 1.9 km

**30** Dose for child
= (Dose for adult × mass of child) ÷ 70
If the dose for an adult is 15 mL and the mass of the child is 28 kg, then
Dose for child = (15 × 28) ÷ 70
= 420 ÷ 70
= 6
The child's dose is 6 mL.

**31** The volume of a rectangular prism is found by multiplying the length by the breadth by the height.

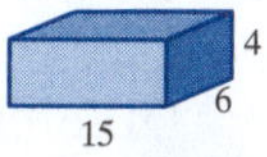


$V = 15 \times 6 \times 4$
$= 360\ \text{cm}^3$

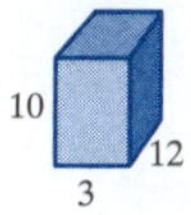


$V = 3 \times 12 \times 10$
$= 360\ \text{cm}^3$

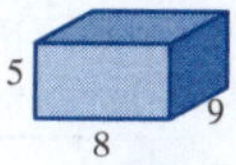


$V = 8 \times 9 \times 5$
$= 360\ \text{cm}^3$

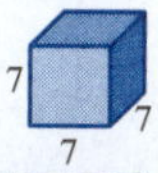


$V = 7 \times 7 \times 7$
$= 343\ \text{cm}^3$

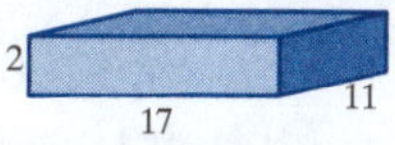


$V = 17 \times 11 \times 2$
$= 374\ \text{cm}^3$

So prisms A, B and C all have the same volume.
[It is not necessary to actually calculate the volumes. The dimensions of prisms D and E are all prime values whereas the dimensions of prisms A, B and C all have the same factors: 2 × 2 × 2 × 3 × 3 × 5]

**32** 4 sandwiches and a drink cost $7.00.
The drink cost $1.80.
So the cost of 4 sandwiches = $7.00 − $1.80
= $5.20
The cost of 1 sandwich = $5.20 ÷ 4
= $1.30
The cost of 3 sandwiches = 3 × $1.30
= $3.90
The cost of 2 drinks = 2 × $1.80
= $3.60
Total cost = $3.90 + $3.60
= $7.50

**33** Find the area of each paddock.
Area = length × width

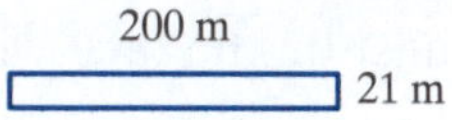


Area = 200 m × 21 m
$= 4200\ \text{m}^2$

Option B:

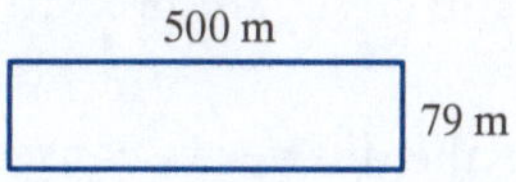


Area = 500 m × 79 m
$= 39\ 500\ \text{m}^2$

Option C:

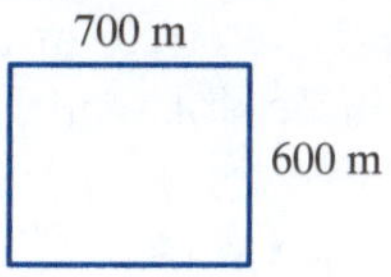


Area = 700 m × 600 m
$= 420\,000\ \text{m}^2$

Option D:

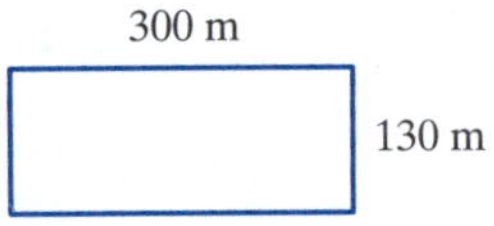


Area = 300 m × 130 m
= 39 000 $m^2$

Option E:

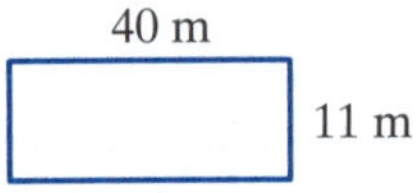


Area = 40 m × 11 m
= 440 $m^2$

Now 1 hectare is 10 000 $m^2$.
So 4 ha is 40 000 $m^2$.
The paddocks that are around 4 ha are options B and D.

**34** Ten 10-cent coins make $1.
So in $17.50 there will be 175 coins.
Length of the line = 175 × 24 mm
= 4200 mm
= 4.2 m

**35** In both views of the cube, a total of 5 faces can be seen.

The face opposite the face showing + can't be seen in either view so it must be the face with the large dot.

So Option C cannot be the answer.

In the first view of the cube you can see that the small dot and the circle are next to each other.

So Option B cannot be the answer because it has the small dot and the circle on opposite faces.

Option D cannot be correct because if the small dot was at the front and the circle at the side, as in the first view, the large dot would be on the top face, not +.

The correct option is A.

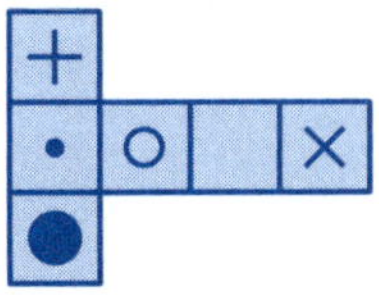

**36** The total men surveyed = 15
So men who support Crows = 15 − (7 + 5)
= 15 − 12
= 3
Total Crows supporters = 3 + 8 + 5
= 16
The total of all supporters = 50
So total Hawks supporters = 50 − (16 + 14)
= 50 − 30
= 20
Children who support Hawks = 20 − (5 + 6)
= 20 − 11
= 9

| | Crows | Eagles | Hawks | Total |
|---|---|---|---|---|
| Men | **3** | 7 | 5 | 15 |
| Women | 8 | **4** | 6 | **18** |
| Children | 5 | 3 | **9** | **17** |
| Total | **16** | 14 | **20** | 50 |

**37** The spinner has eight sectors, three showing 1, two showing 2 and two showing 3, and one showing 4.

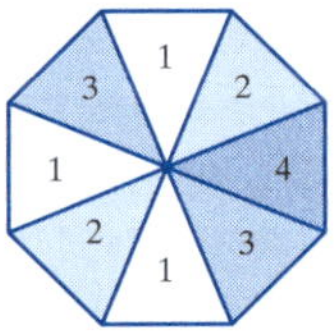


So in every 8 spins you would expect to get three 1s, two 2s, two 3s and one 4.
Now 40 = 5 × 8
So in 40 spins you would expect to get 5 times each of those results.
You would expect fifteen 1s, ten 2s, ten 3s and five 4s.
The correct option is C.

**38** The capacity of the large size is halfway between 1.5 litres and 4.5 litres.

Capacity of large size $= \frac{1.5 + 4.5}{2}$ L
$= \frac{6}{2}$ L
= 3 L

Capacity of small size = 600 mL
= 0.6 L

Difference = (3 − 0.6) L
= 2.4 L

The capacity of the large size is 2.4 litres more than that of the small size.

# Year 7 Numeracy Sample Online-style Test Answers

**39** There were 5 men for every 3 women.
So out of every 8 people, 5 were men.
Now $5 \times 8 = 40$ so there were 5 lots of 8 people attending the conference.

Number of men $= 5 \times 5$
$= 25$

Number of women $= 3 \times 5$
$= 15$

New number of men $= 25 + 15$
$= 40$

New number of women $= 15 + 5$
$= 20$

New ratio of men to women $= 40$ to $20$
$= 2$ to $1$

**40** Number of cubes $= 4 \times 3 \times 2$
$= 24$

Now 2 in the middle of the top layer and 2 in the middle of the bottom layer will have just one face painted. All the rest will have more than one face painted.

Number with one face painted $= 4$

Fraction with one face painted $= \frac{4}{24}$
$= \frac{1}{6}$

**41** Consider each option:

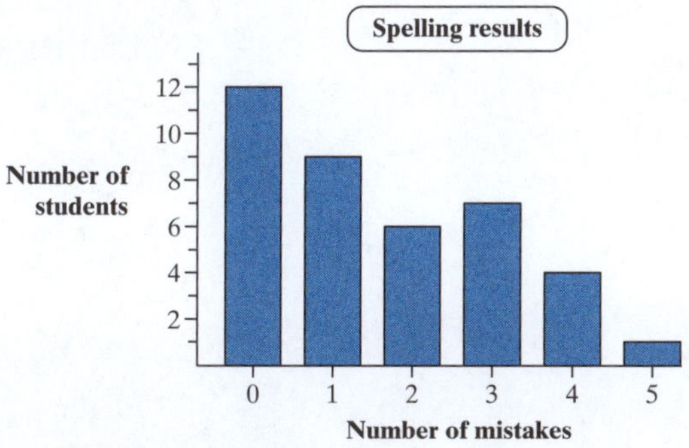


'18 contestants made more than 2 mistakes.'
Number of people making more than 2 mistakes
$= 7 + 4 + 1$
$= 12$
This statement is not correct.

'Altogether 27 mistakes were made.'
Number of mistakes
$= 9 + 6 \times 2 + 7 \times 3 + 4 \times 4 + 5$
$= 63$
This statement is not correct.

'There were 39 contestants.'
Number of contestants
$= 12 + 9 + 6 + 7 + 4 + 1$
$= 39$
This statement is correct.

'No contestant made more than 5 mistakes.'
The graph includes all the contestants and the maximum number of mistakes shown is 5.
So this statement is correct.

The correct statements are 'There were 39 contestants' and 'No contestant made more than 5 mistakes'.

**42** [Write all the fractions with the same denominator (24).]

$\frac{1}{3} = \frac{8}{24}$ [after multiplying numerator and denominator by 8]

$\frac{3}{8} = \frac{9}{24}$ [after multiplying numerator and denominator by 3]

$\frac{5}{12} = \frac{10}{24}$ [after multiplying numerator and denominator by 2]

$\frac{7}{24}$

So, in order from lowest to highest, the fractions are

$\frac{7}{24}, \frac{8}{24}, \frac{9}{24}$ and $\frac{10}{24}$ or $\frac{7}{24}, \frac{1}{3}, \frac{3}{8}$ and $\frac{5}{12}$.

**43**

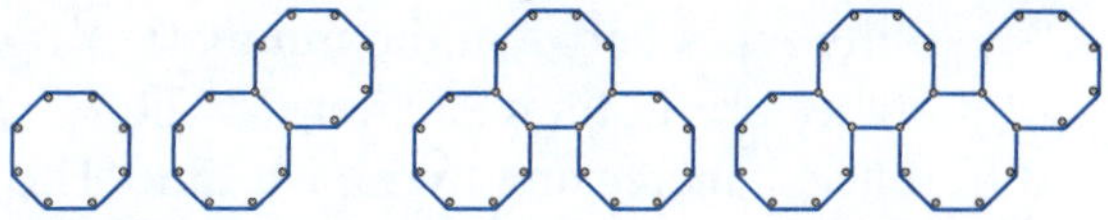

| Number of octagons | 1 | 2 | 3 | 4 |
|---|---|---|---|---|
| Number of pins | 8 | 14 | 20 | 26 |

The pattern for the number of pins is 8, 14, 20, 26, …. The numbers are increasing by 6 each time. So try multiplying the number of octagons by 6.

The number of pins is 2 more than six times the number of octagons.

So to find the number of octagons that use 200 pins first subtract 2.

$200 - 2 = 198$

Then divide by 6.

$198 \div 6 = 33$

So 33 octagons will use 200 pins.

# Year 7 Numeracy Sample Online-style Test Answers

**44** Each angle of a square = 90°
Each angle of a regular hexagon = 720° ÷ 6
Angles at a point add to 360°.
So $x + 720 \div 6 + 90 = 360$
$x = 360 - 90 - 720 \div 6$
$= 270 - 720 \div 6$

**45** Number of cans in bottom row = 6 × 6
= 36
The second bottom row will have 5 cans each way.
Number of cans in second bottom row
= 5 × 5
= 25
There will be 4 × 4 or 16 cans in the next row.
There will be 3 × 3 or 9 cans in the next row.
There will be 2 × 2 or 4 cans in the next row.
There will be just 1 can in the top row.
Total cans = 36 + 25 + 16 + 9 + 4 + 1
= 91

**46**

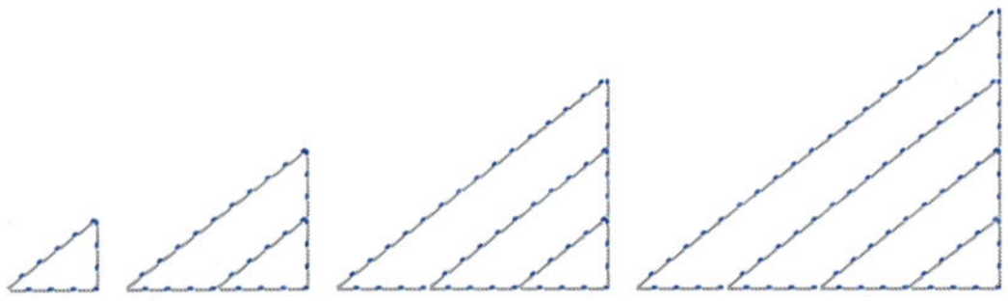

| Figure | 1 | 2 | 3 | 4 | 5 |
|---|---|---|---|---|---|
| Matches | 12 | 29 | 51 | 78 | ? |

The differences between the numbers of matches are 17, then 22, then 27. The difference is increasing by 5 each time. The next difference will be 32.
The next number of matches = 78 + 32
= 110
So the number of matches needed for figure 5 is 110.

**47** The angle for Akmal is a right angle so Akmal received a quarter of the votes.

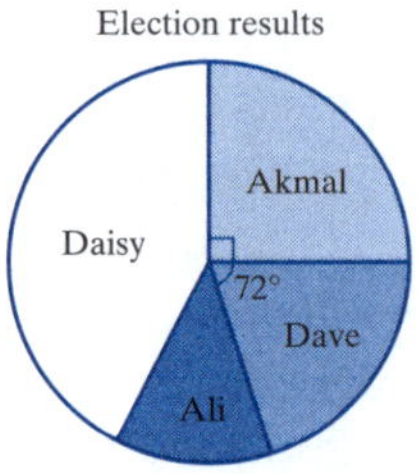


Total number of votes
= 240 × 4
= 960
Dave got twice as many votes as Ali so the angle for Ali is half of 72° or 36°.
Now 3 × 36° ≠ 90°
So Akmal did not get exactly 3 times as many votes as Ali.
But 10 × 36° = 360°
So Ali did get one-tenth of all the votes.
Angle for Daisy = (360 − 90 − 72 − 36)°
= 162°
Ali received 960 ÷ 10 or 96 votes.
Dave received 2 × 96 votes or 192 votes.
So Dave did not get 202 votes.
The correct statements are 'There were 960 votes altogether', 'Ali got one-tenth of all the votes' and 'The angle size for Daisy was 162°'.

**48** Distance from Woodville to Hamlet
= (396 − 258) km
= 138 km
The distance from Woodville to Linden is twice the distance from Linden to Hamlet.
So the distance from Linden to Hamlet is one-third of the distance from Woodville to Hamlet.
Distance from Linden to Hamlet
= 138 km ÷ 3
= 46 km
Now Alice is 396 km from Hamlet.
Linden is 46 km closer to Alice.
Distance to Linden = (396 − 46) km
= 350 km

# Notes

# Notes